BRIDGES

David Blockley is an Emeritus Professor and Senior Research Fellow at the University of Bristol. He has been Head of the Department of Civil Engineering and Dean of the Faculty of Engineering at the University of Bristol. He is a Fellow of the Royal Academy of Engineering, of the Institution of Civil Engineers, of the Institution of Structural Engineers, and the Royal Society of Arts. He was President of the Institution of Structural Engineers in 2001/2002. He has written five other books including *The Penguin Dictionary of Civil Engineering* (2005) and *A Very Short Introduction to Engineering* (OUP, 2012).

BRIDGES

The Science and Art of the
World's Most Inspiring Structures

DAVID BLOCKLEY

OXFORD
UNIVERSITY PRESS

Great Clarendon Street, Oxford, OX2 6DP,
United Kingdom

Oxford University Press is a department of the University of Oxford.
It furthers the University's objective of excellence in research, scholarship,
and education by publishing worldwide. Oxford is a registered trade mark of
Oxford University Press in the UK and in certain other countries

First published 2010
First published in paperback 2012

Published in the United States of America by Oxford University Press
198 Madison Avenue, New York, NY 10016, United States of America

British Library Cataloguing in Publication Data
Data available

Library of Congress Control Number: 2009942577

ISBN 978-0-19-964572-5

Contents

List of Illustrations

LIST OF ILLUSTRATIONS

LIST OF ILLUSTRATIONS

Acknowledgements

Very many people have helped me prepare this book. Firstly I owe an enormous debt to Joanna Allsop, who made many suggestions to make the book accessible to non-technical readers. She read every word and pointed me to the material on Michelangelo's bridge building in the Sistine Chapel.

Robert Gregory and Mike Barnes read the whole book and also made many helpful comments. Ian Firth, John Macdonald, and Jolyon Gill read Chapter 5 and advised me on the dynamics of footbridges. Pat Dallard, Michael Willford, and Roger Ridsdill-Smith read and made helpful suggestions to ground my account of the problems with the London Millennium Bridge in the experience of those who took part. David Weston has provided information about the Bradford on Avon Arch Bridge.

I thank Robert and Ros Gregory for being such good and remarkably patient travelling companions on our bridge-photographing tours in Europe—the photographs Figures 13, 33, 34, and 43 were taken by Robert. I thank Ian May for Figure 4, David Elms for Figure 31, Timothy Bailey for Figure 14, and Mitsuyuki Hashimoto and Dr Hisato Kato for Figure 42. My thanks to Leonardo Fernández Troyano who kindly allowed me to use his picture of the Brooklyn Bridge (Figure 37). David Nethercot, Alistair Walker, and Peter Lewis helped me interpret the evidence regarding the failure of the Dee Bridge. Colin Brown has long

ACKNOWLEDGEMENTS

been a personal mentor and provided material regarding floating bridges in Washington State, USA. Michael Liversidge very kindly and helpfully commented on my attempts to categorize bridges as art and on my sketch for Figure 6, which is based on very limited published material. Richard Buxton suggested I might find material in Herodotus and I did. Patricia Rogers tracked down so much material for me in the University of Bristol Library. Thank you also to the many others who helped directly or indirectly through their conversations, particularly Joan Ramon Casas, Priyan Dias, David Elms, David Harvey, Lorenzo Van Wijk, Guido Renda, Albert Bernardini, Arturo Bignoli, Bob McKittrick, Adam Crewe, Jitendra Agarwal, Mike Shears, Malcolm Fletcher, and Michael Dickson. I send a special word of thanks to Roy Severn and Patrick and Trudie Godfrey for their direct help and encouragement.

I acknowledge permission from the Uffizzi Gallery Florence to use Figure 5; the photo archives of Andreas Kessler, Igis, Switzerland for Figure 8; the Institution of Civil Engineers for Figure 30; and the American Society of Civil Engineers for allowing me to base my drawings for Figure 38 on the diagrams in *ASCE Proceedings* 72 (1946).

I thank the following for permission to quotes extracts: *Burlington Magazine* for text by John Beldon Scott; Penguin Books for material from Herodotus, *The Histories*; the Institution of Structural Engineers for text from the Presidential address by Oleg Kerensky; Simon Caulkin for material from an article in the *Observer* newspaper; and Michael E McIntyre for text from his website.

Particular thanks to David Doran for encouraging me to write this book and, through him, to Keith Whittles. Thanks also to Emma Marchant and Fiona Vlemmiks at Oxford University Press

ACKNOWLEDGEMENTS

and copy editors Charles Lauder Jr. and Paul Beverley. However, the biggest thanks of all go to OUP editor Latha Menon, firstly for having faith in me when she saw my first draft and then for guiding and eventually commissioning the book and for helping to develop it into something worth publishing.

Last, but by no means least, thanks to my wife, Karen, for her long suffering during the gestation, writing, and production of the book and her unfailing love and support and endless cups of tea.

ACKNOWLEDGMENTS

Introduction

Bridges touch all our lives—every day we are likely to cross or go under a bridge. But how many of us stop to consider how the bridge works and what sort of people designed and built it? In this book we will explore how we can read a bridge like a book, to understand how it works, and to appreciate its aesthetic, social, and engineering value.

There are three practical requirements for a successful bridge—firm foundations, strong structure, and effective working. These will form the 'chapters' within which we will find 'paragraphs, sentences, words, and letters'. The 'grammar' of how bridges are put together will be based on combinations of four substructural types—BATS—beams, arches, trusses, and suspensions. For example, the Golden Gate Bridge is a suspension Bridge with a roadway deck on a stiff truss beam.

Bridges are icons for whole cities—think of New York's Brooklyn Bridge, Sydney's Harbour Bridge, and Brunel's Clifton Bridge in Bristol where I live. Traditionally architects have not been involved in bridge design because bridges have been conceived as 'raw' engineered structure. Yet bridges are also a form of functional public art—they can delight or be an eyesore. Now architects and sculptors can and do contribute to the aesthetics of bridges to improve their impact in our public spaces. One of the finest examples is the Millau Viaduct in France which certainly has the 'wow' factor.

INTRODUCTION

Bridge building is a magnificent example of the practical and everyday use of science. Unfortunately there are always gaps between what we know, what we do, and why things go wrong. Bridge engineers must manage risks carefully. They know that information has a 'pedigree' which they must understand. The rare cases of bridge failures can help us to learn some valuable lessons that also apply to other walks of life. One example is that failure conditions can incubate over long periods and we can learn to spot them. Another is that partial or 'silo' thinking with a lack of 'joined-up' inadequate processes do typify technical and organizational failures.

The first chapter focuses on why bridges are important and set out the basic BATS grammar that we will use to read them. Chapters 2–5 describe how arches, beams, trusses, and suspension cables work, using real and specific examples. Arches are symbols of stability. Beams bend flexibly. Trusses are physical teamwork. Suspension bridges are often landmark structures. Chapter 6 sets out the role of scientific models and managing risks and addresses the question 'How safe is safe enough?'

The final chapter shows how pragmatic systems thinking is natural for bridge builders. They use practical rigour which is not the same as scientific logical rigour. Logical rigour is necessary but not sufficient for practical success. Bridge engineers also build people bridges as they form and reform teams to accomplish their successes. We will synthesize the lessons from bridge building and show how they feed into problems where 'joined-up' thinking is needed.

There are many books published about bridges. This book is different in that it is a rare, indeed the only example as far as I am aware, attempt to help nontechnical readers understand the technical issues that bridge builders have to face.

1

BRIDGES ARE BATS
Why We Build Bridges

'Architects are a strange breed,' wrote BBC's Andrew Walker in 2002. He was referring to British architect Lord Foster who, according to Walker, designed the famously 'wobbly' London Millennium Bridge. 'No other profession stamps its personal style on our lives in the way that theirs does,' he said.[1] Jonathan Duffy, also from the BBC, writing in 2000 had anticipated his colleague's comments. 'As the Millennium Bridge shows, modern architecture is anything but a breeze. At the cutting edge, uncertainty is an occupational hazard.' He continued: 'It [the bridge] was supposed to be a blade of light shooting across the Thames.'[2] For the 'hordes' crossing at the opening 'the experience was more like a rickety fairground ride.'

Nowhere in the article did Duffy mention the real technical designers of the bridge who did most of the work—the structural engineers, Arup. Indeed he attributed the complex calculations and computer models to Foster when in fact Arup had done them and got them right—but the bridge still wobbled.

So what went wrong? Should Arup be blamed for designing a bridge that manifestly didn't do its job? Innovation is always risky but without risk we don't advance. Did Arup take too many risks in a single leap?

Arup is one of the leading firms of structural engineers in the world. They knew that when they had a wobbly bridge, they just had to put it right—and they did so. It's now as steady as a rock and a real landmark.

The reason for the wobble is interesting. I was part of the crowd invited to test the bridge before it was reopened in 2002. Several hundred of us processed back and forth as its movements were measured. Arup had fitted shock absorbers and wanted to check that they would prevent the wobbles—and they did.

Afterwards I took a taxi back to the Institution of Structural Engineers HQ where I was staying as President for 2001–2. I told my taxi driver what I had been doing. He said, 'Typical ain't it? We Brits can't even get a bridge right.' As a structural engineer I felt defensive. I told him that the wobble was a different kind of wobble to the well-known vibrations created by soldiers marching. The engineers knew that soldiers marching in step caused vibrations and had correctly done extensive calculations for that kind of wobble. The problem was that this wobble was a sideways one that hadn't really been recognized before. 'It's called synchronous lateral excitation', I said. He didn't ask me to explain anymore because by then we had arrived at my destination.

The story of the wobbly bridge that no longer wobbles is a classic example of real progress in knowledge being made arising from something unforeseen going wrong.

In 1996 the *Financial Times* newspaper and Southwark Council organized a competition to design a new footbridge across the Thames. The winning team was Arup, Foster and Partners with sculptor Sir Anthony Caro. Their design decisions were con-trolled by the architectural vision of the bridge as a 'blade of

FIG 1. London Millennium Bridge

light'—a vision to which all members of the team contributed. A suspension bridge was, from a structural point of view, unlikely to be the most economic solution but it suited the overall concept. The team chose an unusual form of shallow suspension bridge where the tensioning cables are, as far as possible, below the deck level so that all views were unobstructed. In fact the cables sag around six times less than those of a conventional suspension bridge. The team decided on three main spans of 81, 144, and 108 m from north to south (Figure 1). The bridge deck was designed and built with steel box arms spanning between the cables every 8 m. The deck structure has two steel edge tubes supported by the arms and the 4-m-wide deck is of aluminium. Two piers were built to support the bridge from the river bed and the eight suspension cables pull against the abutments set into each bank with enough force to support 5,000 people on the bridge.

The shallowness of the cables means that the pulling tensions in the cables are higher than normal, making the bridge taut and highly strung. They act rather like the strings of a violin. When a violinist tightens a violin string to make the note higher that makes it vibrate at a higher frequency so that the strings move backwards and forwards through repeated cycles at a higher rate.

All bridges and other structures, including the human body, have what scientists call a natural frequency—when objects vibrate freely. Bridges with spans similar to the London Millennium Bridge typically vibrate with natural frequencies between 0.5 and 1.0 cycles per second. If wind or pedestrians apply forces to the bridge at the same frequency as the natural frequency then resonance occurs and the vibrations can become very large indeed. When we walk across a bridge we push down with each step but we also push outwards slightly as well. The structural engineers did extensive calculations and thought that they had all these possible sources of wobble covered.

The bridge opened on 10 June 2000. It was a fine day and the bridge was on the route of a major charity walk. There were around 90,000 users on that first day with up to 2,000 on the bridge at any one time. The bridge swayed from side to side unexpectedly and was closed two days later. It was dubbed the 'wobbly' bridge by the media who declared it another high-profile British Millennium Project failure. Not everyone agreed—some people were reported as saying that they enjoyed the swaying around and one even said it was a shame the bridge wasn't more wobbly.

So what do engineers do in the face of such a public problem? Arup decided to tackle the issue head on. They immediately undertook a fast-track research project to seek the cause and the cure. Measurements were made in university laboratories of the effects of people walking on swaying platforms. Large-scale experiments with crowds of pedestrians were made on the bridge itself. From all of this work, involving a number of people and organizations, a new understanding and a new theory were developed.

The unexpected motion was the result of a natural human reaction to small lateral movements. If we walk on a swaying surface we tend to compensate and stabilize ourselves by spreading our legs further apart—but this increases the lateral push. Pat Dallard, the engineer at Arup who was a leading member of the team who developed the new theory, says that you change the way you walk to match what the bridge is doing. It's rather like walking on a rolling ship deck—you move one way and then the other to compensate for the roll. The way people walk doesn't have to match exactly the natural frequency of the bridge as in resonance—the interaction is more subtle. As the bridge moves, people adjust the way they walk in their own manner. The problem is that when there are enough people on the bridge the total sideways push can overcome the bridge's ability to absorb it. The movement becomes excessive and continues to increase until people begin to have difficulty in walking—they may even have to hold on to the balustrades. One of the difficulties is that there is no sign of any trouble until a 'critical number' of pedestrians are on the bridge. In tests on one span of the London Millennium Bridge, there was no sway at all with 156 walkers on it, but when 10 more walked on it a wobble started and increased rapidly.

The intense media publicity brought to light some previous eyewitness accounts of this kind of wobble. Examples in the 1970s included UK bridges at the National Exhibition Centre and in Chester and also the Auckland Harbour Road Bridge, New Zealand, during a Maori demonstration. One month after the Millennium Bridge opened a 100-year-old road bridge in Ottawa wobbled as a huge crowd left the bridge. The Golden Gate Bridge on the day of its opening and the Brooklyn Bridge during a power outage have both also suffered. A colleague at the University of Bristol, John Macdonald, has recently measured similar movements on Brunel's Clifton Suspension Bridge in Bristol.

The only documented technical study before the millennium celebrations was in 1993 by a Japanese team lead by Yozo Fujino of the University of Tokyo. In a technical research paper they wrote, 'It seems that human-induced lateral vibration has not been checked in designing pedestrian bridges.'[3] They reported experiments on a cable-stayed pedestrian bridge next to a boat race stadium. After a race as many as 20,000 people passed over the bridge in 20 minutes. The Japanese team produced evidence demonstrating synchronized walking and lateral vibration of the bridge. Unfortunately the paper was published in a research journal about earthquake engineering rather than one directly concerned with bridges—an illustration of the difficulty of sharing this kind of information.

The solution to stop the wobble of the London Millennium Bridge was to install shock absorbers, rather like in a car. Using the results of their quantitative research the engineers designed a system of 37 shock absorbers called 'viscous dampers' and 54 weights attached to the bridge by springs to dampen the vertical motion. The research and design process took over four months.

The actual work cost over £5m and the bridge was reopened on 22 February 2002. With some style, Arup organized an opening concert and commissioned a special piece of music for the occasion called Crossing Kings Reach by Peter Maxwell Davies.

So, the particular wobbles of the wobbly bridge were not anticipated by the engineering designers. They had missed the Japanese research. Although the phenomenon had been seen before by a few researchers it hadn't been recognized sufficiently as something that bridge designers should be looking for and it had not found a place in any bridge design codes, manuals, or journals. The phenomenon was rare because the susceptible bridges had not experienced the critical number of pedestrians. The problem was that there is nothing to see until you get a big crowd—and that may simply not happen.

Bridge builders now realize that potentially this can happen to any long bridge carrying pedestrians. According to the latest theory the 'critical number' of people above which these wobbles will occur depends on the weight of the bridge, its natural frequency, and the amount of damping (i.e. the degree to which the bridge has 'built-in' shock absorbers). Larger bridges are more like double basses than violins and so will have lower natural frequencies. Many bridges will be heavier than the Millennium Bridge though the level of damping will vary, depending on the individual design.

This means that the critical number of people to make a given bridge wobble will usually be larger than was originally the case for the Millennium Bridge. The 'cure' was to increase the damping to a level where the critical number of people is more than can reasonably actually get onto the bridge.

Although these kinds of wobbles can occur on any long bridge they were, and still are, very rare. Arup's design did not cause the wobble because of its innovative structural form; the wobble arose as a result of the large number of people and insufficient damping in the structure. So the popularity of the bridge on its opening day put a spotlight on the susceptibility of all bridges. As a result the phenomenon has now been researched to a point that future bridge designers will be able to take it into account. Certainly, bridge owners need to take advice if there is a chance that very large numbers of people might congregate on their bridge.

So were Arup to blame? The simple answer is 'no'. Arup were responsible but not blameworthy—an important distinction. They followed best practice but best practice was not good enough. When designers innovate there is a need to take great care in checking for new and, unintended consequences. Bridge building is a risky business and, as we will discover in Chapter 6, the risk of unintended and unwanted events is always present.

One could argue that the whole issue of the wobbly bridge was cultural and not technical. The problem might have been avoided if people's expectations had been managed differently. There are bridges across the world that do wobble a great deal but in those cases pedestrians are warned before they cross and so they know what to expect. The wobbles of the Millennium Bridge would not have caused it to collapse (although its life may have been curtailed due to metal fatigue) so there was no threat to life. If the bridge had been designed expecting it to wobble and people were warned of the possibility then all of the fuss made wouldn't have happened. There might well have been complaints but as we shall see in Chapter 5, the Capilano Bridge north of Vancouver is very wobbly and the Carrick-a-Rede rope bridge in County Antrim, Northern Ireland, is said to be so bouncy it's a tourist's challenge! Forewarned is forearmed.

The bridge as a book

We are going to explore how to read a bridge like a book. As we do so the story of bridges that will evolve has many interwoven strands of artistic, technical, scientific, and cultural development. As we sift out the letters, words, sentences, paragraphs, and chapters of the book of a bridge and delve into the grammar of bridge structures we will begin to appreciate their aesthetic, historic, social, and engineering value.

Bridges aren't just built to cross obstructions; they help us express some of our deepest emotions. The London Millennium Bridge is just one example of building as a way of commemorating a significant anniversary. All through history people have expressed their awe, wonder, spirituality, and religious faith by building. Pyramids were a connection, a bridge, between this world and the next. Churches and cathedrals contain soaring arches to reach out to the heavens and to bridge the roof. Even when we want to express naked power we build structures—the old medieval castles, with drawbridges, are examples. Modern skyscrapers serve to demonstrate the economic power of multinational companies. Of course a building is not a bridge but buildings are full of small beams bridging over the spaces below. We won't be considering buildings in any detail in this book but it is worth noting that the floors in some buildings can span over very large openings such as the ground floor foyer of a large office block or departmental shopping store. Even at home the timber trusses in the roof of your house bridge over the space where you live and the lintel over the door or window is a small bridge.

9

Bridges can be delightful or disagreeable to look at. They can be a form of public art or a functional eyesore. London's Tower Bridge, New York's Brooklyn Bridge, and Brunel's Clifton Bridge in Bristol are icons known and recognized throughout the world. Television pictures of fireworks on Sydney Harbour Bridge are beamed around the world to herald in the New Year. Such traditional bridges are 'raw' engineered structures with little architectural or sculptural involvement, yet architects and sculptors can and do contribute to the aesthetics of bridges—and more so recently. The effect is to improve their impact on our public spaces. The final outcome is a real team effort involving many different forms of creativity.

Bridges are links; they connect people and communities. They enable the flow of people, traffic, trains, water, oil, and many other goods and materials. Bridges therefore contribute to our personal well-being and our quality of life. They can help whole regions to develop socially and economically.

Bridge building is an art and a science. Bridge builders use science but they are not applied scientists. This is because there are always gaps between what we know, what we do, and why things go wrong. So bridge engineers must learn to manage risks carefully. The rare cases of bridge failure can help us to learn some valuable lessons that apply to other walks of life. One important lesson is that a lack of 'joined-up' thinking typifies technical and organizational failure. Our story will therefore also include examples of what can go wrong—sometimes resulting in the dramatic collapse of a complete bridge. We will explore some of the lessons that have been learned.

The London Millennium Bridge is one of the latest bridges to be built over the Thames. In Chapter 2 we will look at the first London Bridge which also had problems—so many so that they inspired the nursery rhyme 'London Bridge is falling down'.

Ancient bridges upset the river gods and had to be placated, often with human sacrifice. From river fords and stepping stones to the first bridges of simple tree trunks and stone slabs; from the Forth Railway Bridge to the Millau Viaduct in France, the story of bridges is as much the story of the people who built them.

There are three practical requirements for a successful bridge—firm foundations, strong structure, and effective working. Firm foundations are especially critical for traditional structures such as arches. Indeed once erected, arches will stay in place for a very long time as long as the foundations don't move. All bridges require strong robust and stable structure. However, the real test for a successful bridge is whether it works effectively. Bridges stand up because the basic structural components interact and work effectively with each other. The foundations, strength, and effectiveness of the aesthetic, social, and cultural aspects of bridges are much more difficult to capture but are nonetheless very important.

Bridges are described in many different ways. If you were to attempt to capture all of the types mentioned on the Internet your list would be very long and confusing. To begin to read a bridge we need some principles to help us classify them.

It is helpful to start by thinking of bridges from three different perspectives—*purpose, material,* and *form.* The purpose of a bridge is the first and most basic requirement. It embeds the bridge in its technical, social, cultural, and historical context. A purpose defined without recognizing all of these requirements will be partial. A strong but ugly bridge is inadequate. Worse is a weak but beautiful bridge because strength is a necessary requirement although it is not sufficient. A high-quality bridge is one that is 'fit for purpose' but this is true only if all aspects, all angles and points of view, including affordability and sustainability, are appropriately specified.

The purpose will specify how the bridge will be used; it will strongly influence the form of the structure, the materials it will be made from, and how it will be erected. For example, a bridge over navigable water must allow ships to pass—so some bridges may have to lift or swing. A bascule bridge operates like a seesaw usually with a big weight balancing the rising deck of the bridge. However, for most bridges the main purpose is reasonably obvious and simply captured. Footbridges, highway bridges, and railway bridges carry pedestrians, road traffic, and trains—nothing very complicated about that—except that different structural solutions may be required for spanning over rivers, railways, roads, or deep valleys.

The list of materials from which bridges are made is actually quite short. It includes timber, masonry, concrete, iron, steel, and more recently aluminium and plastics, but little else. Bridge materials must be strong enough for the job they will be asked to do, readily available, and not too expensive. Of course combinations of material are used. For example, because concrete is strong when squashed but weak when pulled, steel bars are used to reinforce it.

The choice of the form of a structure is one of the most critical decisions that a bridge builder must make and it is the focus of much of this book. First and foremost the structure must be able to stand firm whatever happens and so unsurprisingly that is a major preoccupation. Whatever the natural or man-made hazards, the bridge must be safe. High winds, heavy rain, earthquakes and tidal waves, very heavy lorries and trucks, and even terrorist attacks have to be resisted.

A little later in this chapter we will classify structural form using combinations of BATS—beams, arches, trusses, and suspensions. However, because the strength of a bridge is so crucial it's worth first considering the three ways in which materials are strong—pulling, pushing, and sliding. Scientists and engineers use the term tension for pulling, compression for pushing, and shear for sliding. These three ways to be strong are expressed in BATS in different ways—so let's look now at each one in turn.

Tension

Imagine a tug of war between two teams with say five people in each team. Each team is pulling on a fairly substantial rope and there is a tag on the rope right in the middle. The referee of the contest watches the tag because the team that pulls it towards them a measured distance will win. Imagine that we are looking at the rope at the moment when both teams are pulling equally hard so the result as to which team will win is in balance—the tag on the middle of the rope is not moving either way.

We want to understand the strength of the rope in tension— being pulled. So let's think about what is happening inside the rope at that point where the tag is attached. One way to do this is to carry out a thought experiment—in other words, to mentally do something to the rope and think what would happen as a consequence. So what we'll do is imagine that we can cut the rope at the tag and separate the two halves of the rope. What would happen? Both teams would collapse in a heap! They would suddenly be pulling against nothing—just as if the rope had snapped.

So to prevent our teams from falling we would have to get the two halves back together and replace what the internal fibres were doing before we cut the rope. To do that we would have to pull with a force equal to that produced by the two teams in both directions. We would have to pull against one team one way and against the other team the other way at the same time. Imagine

13

doing that yourself: you would have to get hold of both cut ends and pull them towards you to balance the pull of both teams.

The force that you are now providing as a substitute for the fibres of the rope is called an *internal* force. This internal force is a response to the *external* force from the teams. This distinction between internal and external forces is essential to an understanding of the way bridges work and we will constantly be referring to it throughout our story. When the internal forces balance the external forces the rope is said to be in *equilibrium*—everything just balances out. If the teams pull so hard that the internal force gets so large that you have to let go (or else your arms will be pulled out of their sockets) then their pull defines the breaking strength. Of course that's *your* breaking strength. You could find the real breaking strength of the rope by pulling it until the internal force gets so large that the fibres snap. The rope will be too strong for you to do this manually but you could do the same thing with a piece of cotton. In reality engineers and scientists use a special testing machine in a laboratory to apply varying tensions large enough to break lengths of rope and pieces of steel or other materials used in a real bridge to find out how strong they are.

The internal force is acting all along the length of the rope from one of the teams to the other. We could have made our cut anywhere along its length and used the same argument. So we call the force an axial tension—it is acting axially along the length of the rope. The cross section of the rope is an end view of the cut. The area of the cross section of your rope is quite small. In a real bridge with lengths of steel or timber in tension the area of a cross section will be much larger and the internal force may not be exactly along the axis of the member. As we shall see a little later, the action of the axial force can also be described by saying that the rope has just one 'degree of freedom'—in other words, just one way of changing.

Force is measured in newtons (usually abbreviated to N).[4] The tension might not act exactly along the axis of the rope so it is generally better to consider the force on each little element of the cross section. The force on a small element is called a stress and using it we can consider how stress varies across the cross section. Consider a rod with a square cross section which is 10 mm by 15 mm, which therefore has an area of 150 mm². Imagine the rod is pulled by an axial force of 15,000 N (or 15 kilonewtons = 15 kN) so that the stress over the cross section is the same. The stress will then be 15,000/150 = 100 newtons per square millimetre (usually abbreviated to N/mm²) uniformly across the section. This way of expressing an internal force as a stress is another part of a very powerful set of mathematical tools used by bridge builders.

So far we haven't said anything about how much the rope stretches when it is pulled. Imagine that your rope was made from a gigantic elastic band or a length of coiled spring. Clearly when you pull, the band or the spring would stretch quite a lot. In fact all materials stretch when pulled, and some stretch more than others. The amount of stretch is very visible for an elastic band or a spring but it is so very small for a piece of wood or steel that you need a special measuring instrument to detect it. The amount of stretch of a material is crucially important in bridge building because it contributes to two things—how much the bridge will deflect and how much it will vibrate as the wind blows or as heavy lorries pass over it. This stretching is not the only factor in deciding the amount of deflection or vibration but it is an important one. Scientists and engineers are interested in the amount of stretch for every unit length of a piece of material and they call it strain. So if a piece of string 1 m long (i.e. 1,000 mm) stretches by 10 mm the strain is defined as 10/1000 or 0.01. Note that strain has no units—it is dimensionless.

If the amount of stretching is important in bridge design then even more important is the amount of stretching produced by a particular level of force. The amount of force required to create an amount of stretch is called the stiffness of the bar. So if, as before, a force of 15 kN stretches a 1-m bar by 10 mm then the stiffness of the bar is 15/10 or 1.5 kN/mm. This is distinguished from the stiffness of the material which is defined as the amount of stress required to create an amount of strain in that material. That is called the elastic modulus. Thus if a stress of 100 N/mm^2 makes the 1-m length of rope stretch by 10 mm (which as we calculated above is a strain of 0.01) then the stiffness of the material is the stress divided by the strain or 100/0.01 = 10,000 N/mm^2. Note that because the strain is dimensionless then the units of elastic modulus are the same as the units of the stress.[5] For many materials the elastic modulus remains the same for various loads. We can show this by plotting a graph of stress against strain. For many materials the result is a straight line and its slope is the elastic modulus. Such a material is said to be linear elastic.

Compression

Now let's turn our attention to the opposite of tension—the effect of pushing, squashing, or compression. If our tug of war teams were to push on the rope rather than pull on it the rope would just fold—you can't push on the end of a rope—it has no stiffness in compression.

So what can we do? We could decide to replace the rope with a wooden rod or pole and hold a 'push of war' competition. But the pole would have to be quite long. The teams could push on it to some extent but unless the pole was very thick and chunky it would soon buckle and break. Long, thin materials such as rope, string, and long, thin poles are strong in tension but soon buckle in compression. In order to generate the same force in compression as in tension, e.g. two teams of five people all pushing on a wooden rod together, you would need a massively thick piece of timber like a battering ram. Thus we can immediately see that it is much more difficult for a material to resist a pushing force—a compressive force.

Two main factors determine the strength of a rod in compression, its length and the shape of its cross section—Chapter 4 has more detail on this. The way the rod is held at its ends is also influential. The longer the rod, the more likely it is to buckle. A very short rod will not buckle at all—it will just squash. Just imagine standing on a single brick—it can carry a very big load before it squashes by crumbling. Indeed we usually think of a single brick as a rigid block—meaning that the strain is so small before the final crumbling that we can neglect it. This property is used when building masonry arches, as we will see in Chapter 2. Arches are one of the oldest forms of bridge and they rely on materials such as masonry that are strong in compression.

Shear

The last way in which structures must be strong is in shear. Shears, like scissors, are used to cut, so shearing is a cutting or slicing action. In a bridge structure a shear force is a force that resists slicing or sliding.

Think of a block or brick sitting on a relatively rough flat surface and imagine pushing it horizontally. At first there is some resistance but if you push hard enough eventually it slides as you overcome the friction. Now think of two blocks, one on top of the other but we stop the bottom block from moving by putting some kind of solid obstruction in its way and we push against the top block. The top block will slide over the lower block in just the same way. Then replace the two blocks by one new solid block which is made of the same material and is the size of the two blocks together. Again the obstruction prevents the lower part of this new block from moving. When you push against the top part of the new block you are doing exactly the same thing as when there were two blocks—except they are now joined together. The material of the block is holding the two parts together by resisting the tendency to slide. The force required to do this is an internal shear force which acts at the interface between the two previously separate but now joined up blocks. Of course the solid block could be separated into two blocks at many different levels so the same argument can be used to show that a shear force is created at every possible level of the block.

When we were considering tension we defined a force on a small piece or element of the rope as a stress. For shear the situation is a little more complicated. We need to think about a small piece or element of the block, say a small cube with sides of 1 mm. If the cube is in equilibrium the horizontal internal shear force acting on the top of it has to be balanced by a shear force of the same size on the bottom. However, although these two forces may balance each other horizontally the two together would create a tendency for the piece to rotate. So if there is no rotation and if equilibrium is to be maintained then an internal shear force has also to be generated on the vertical faces of the piece—one up and one down. It follows that shear force acts both horizontally and vertically on our small elemental cube as in Figure 2a. If the block were to be subjected to a twisting motion then there would also be shear forces on the other faces of the cube. It's worth just noting that so far we have, perhaps somewhat arbitrarily, just been talking about a cube with horizontal and vertical sides and I will continue to do that for most of the rest of the book. However, there is a set of shear forces acting on any element of any orientation we may care to define within the block.

Just as earlier we replaced the stiff rope with an elastic band or coiled spring in order to make the tensile strain visible so now we will need to replace the block with an elastic rubber block if we want to see some shear strain. If we do that then the top of the block will move visibly compared to the bottom. Looked at from the side the block becomes a lozenge shape as the top moves and the bottom stays still. Consequently one diagonal lengthens and the other shortens; one diagonal is in tension and the other is in compression as shown in Figure 2b.

In Chapter 2 we'll see how shear is important in the way sandy soils carry forces. In Chapter 3 when we look into how a beam bends we will find that there can be a turning effect on our small element when the shear forces change along the length of the beam and this creates another internal force which is called a bending moment. In that case the element will rotate as well as distort into a lozenge shape.

How are these three ways of resisting forces expressed in the forms of a bridge? As I have said, BATS is an acronym for beams that bend, arches that compress, trusses that compress and stretch, and suspension bridges that hang. The chapters, sections,

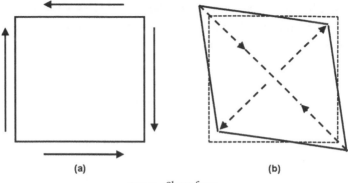

FIG 2. Shear force

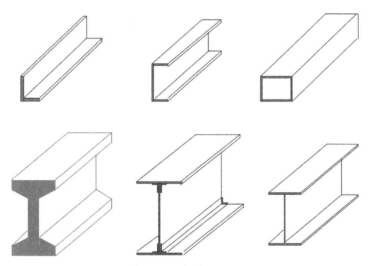

FIG 3. I-beams and other components

and paragraphs of the superstructure (i.e. the structure above ground) of our bridge book are various combinations of BATS. The substructure and foundations are generally, but not always totally, below ground and unseen and they form an important chapter too. As always, the story is not straightforward because most bridges are mixtures. For example, Ironbridge, Coalbrookdale, Shropshire, UK, was the first cast-iron bridge, built in 1779, and has a truss acting as an arch. The decks of all modern suspension bridges, like the Golden Gate Bridge in San Francisco, USA, are beams or trusses. So we will look in some detail at all of these forms and how they are combined in Chapters 2 to 5. We will explore how to read the various ways in which strength in tension, compression, and shear is used to carry forces.

So that's how we'll deal with the chapters, sections, and paragraphs but what about the sentences, words, and letters? The sentences are the individual structural components that can be used in many forms of bridge. Each one must be able to resist internal forces too. Indeed most of them are particularly shaped in order to do that efficiently for one or more types of force. Some common examples are plates, tubes, I-section beams, channels, angle sections, circular and rectangular tubes, wires and cables—Figure 3. As we'll see in Chapter 3, I-section beams are shaped the way they are to be efficient in bending. Cables are strong in tension but can carry no compression just like the rope we discussed earlier. These components may be manufactured in a factory or built on site. For example, steel companies make steel beams by rolling ingots of steel in gigantic presses. They are then transported to a steel fabricating workshop or yard and assembled into parts of the bridge before being taken to the bridge site. Concrete beams, on the other hand, are often cast on site or *in situ*. The sentence components of our bridge book will also include some manufactured assemblies such as bridge bearings—see Chapter 3.

The bridge book's words are the materials from which the sentence components are made. So, for example, it is possible to buy sentences of manufactured beams made from steel, pre-cast concrete, and solid or laminated timber. The various constituents of these materials are the letters. Steel is a familiar but complex manufactured alloy. It is made from iron and carbon with small amounts of other additives such as magnesium. The amounts of carbon and other added metals determine the strength and ductility of the steel and must be carefully controlled during manufacture. The chemical bonds that give a material its strength can change with different treatments and these must be understood by the bridge builders. So, for example, when high strength steel with a high carbon content is welded, precautions must be taken to prevent the steel becoming brittle. Concrete, by contrast, may seem to be a rather simple and commonplace material. However, this familiarity can be misleading. The chemistry of concrete is very complex, which means that concrete made for any structural work must be carefully controlled. For example, the ratio of the amount of water to cement is critical for the strength of concrete and for the ease with which it is cast. Various additives can be used to improve 'workability'—the ease with which the concrete is placed. A considerable amount of heat is produced during the casting and curing of a large volume of concrete and this must be properly controlled if the concrete is not to deteriorate.

So far we have said little about the ways in which the words, sentences, paragraphs, and chapters relate to each other—the grammar of our bridge book. We haven't recognized the subtleties of the layers of meaning in the written word—from novels to poetry—in our bridge book. We need, of course, to know something about the rules at different levels by which the words combine to make sentences and the sentences to make paragraphs and so on. As one might expect, the full and complete grammar is complex. Nonetheless we can read our book of bridges by focusing on each level in turn. We can begin to see just how the structure resists all of the external forces, such as road traffic or wind pressure, to which the bridge is exposed. We can begin to understand the internal forces,[6] which are in equilibrium with the external forces such that they flow through the bridge down to the foundations. As we work with these ideas we can also begin to appreciate some of the subtleties of the layers of meaning in the role of bridges in our working infrastructure. Bridges are not just physical objects; they are embedded in our technical, scientific, aesthetic, social, and cultural heritage.

I mentioned earlier that a rope in tension has one degree of freedom. Let's now look at this idea more closely since it forms one of the first parts of our grammar. In brief, degrees of freedom are the independent directions in which a bridge or any part of a bridge can move or deform. We'll come back to the important word, independent, a little later. Degrees of freedom define the shape and location of any object at a given time. Each part, each piece of a physical bridge whatever its size is a physical object embedded in and connected to other objects. Whether the object is a small element of a cross section $1\,\text{mm} \times 1\,\text{mm}$ or a large substructure, it is connected into other similar objects which I will call its neighbours.[7] Whatever its size, each has the potential to move unless something stops it. Where it may move freely then no internal resisting force is created. However, where it is prevented from moving in any direction a reaction force is created with consequential internal forces in the bridge structure. For example, at a bridge support, where the whole bridge is normally stopped from moving vertically, then an external vertical reaction force develops which must be resisted by a set of internal forces that will depend on the form of the bridge.

So inside the bridge structure each piece, however small or large, will move—but not freely. The neighbouring objects will get in the way, rather like people jostling in a crowd. When this happens internal forces are created as the objects bump up against each other. These forces can be managed by considering how they act along the directions defined by the degrees of freedom. The degrees of freedom represent the directions and the ways in which the structure has to resist internal forces— they define the ways in which the bridge has to be strong.

The degrees of freedom are defined using a mathematical convention. A rigid body like a brick has six degrees of freedom. It can move along three axes at right angles and can rotate about each of them.[8] I said earlier that these directions must be independent. What this means is that it must be possible for a change in one direction to occur without a change in the others. Without independence it would not be possible to separate out the various movements and the consequential internal forces. Nor would it be possible to combine them again. Of course in general they all do change together. By describing those changes in terms of the degrees of freedom the bridge builder can make sure that each object of the bridge is able to resist the internal forces along these six directions. The grammar of science helps bridge engineers understand the nature of these internal forces (for example, whether they are tension, compression, shear, or some combination to create a turning and/or twisting effect) and how to calculate them.

Until the advent of powerful computer programs in the last part of the twentieth century all but three of the degrees of freedom were dealt with by common sense because it was too complicated to calculate them. So, for example, two bridge beams or trusses were placed side by side and braced together. This stabilized them sideways or laterally, and prevented twisting. It was an intuitive and effective way of dealing with all of the degrees of freedom that have a sideways lateral component. As a consequence the focus for many centuries of engineering scientific development was on the other three degrees of freedom. The first was the vertical movement, which contributes to the displacement of the bridge. The second was the horizontal movement along the length of the bridge, which contributes to its change of length. The third was rotation in the vertical plane as you look at the side view or elevation of the bridge, which when resisted creates internal bending moments. In Chapter 6 we will examine how modern engineering science and, in particular, extremely fast computer calculations, have enabled us to develop 'finite element' techniques of theoretical analysis through which we can understand, calculate, and control all six degrees of freedom for parts of the bridge. We shall see that this knowledge is making it possible to build much more exciting and unusual bridge forms.

Forth Railway Bridge

Let's now begin to learn how to spot the paragraphs, sections, and chapters of the bridge book. Alfred Hitchcock's film of *The 39 Steps* (1935) showed many people, for the first time, one of the world's greatest bridges and one of the most famous achievements of Victorian engineering, the Forth Railway Bridge (Figure 4). In the film an innocent man on the run escapes on the

FIG 4. Forth Railway Bridge

bridge and hides somewhat riskily, clinging to the steelwork 150 feet above the icy waters. At this stage I will talk mainly about the superstructure and only briefly mention the foundation substructure which has been working unseen since 1890. One of the main reasons I have chosen it as a first example is because although the bridge looks quite complicated at first, the flow of forces is more transparent than most.

The Forth Railway Bridge is a steel truss structure that has carried trains over the Firth of Forth from just north of Edinburgh to Fife since it was opened on 4 March 1890. The work began in February 1883 to complete the east coast railway route between London and Aberdeen. The Tay Bridge had collapsed in 1879 and this influenced its design (see Chapter 4). In 1964 the rail bridge was joined by a close neighbour—the Forth Suspension Road Bridge. The rail bridge is over a mile long and higher than the dome of St Peter's in Rome. It was the longest railway bridge in the world at that time. It appeared then to be such a rebuff to the natural order that it was condemned by William Morris as the 'supremest specimen of all ugliness'. Immanuel Kant wrote that objects of sublime contemplation and beauty 'raise the forces of the soul above the heights of the vulgar commonplace'—the sight of a steam locomotive thundering across the impressive spans of the Forth Bridge was, to many, a sublime experience.

Even today, if you stand below the bridge and look up high into the girders you feel simply overwhelmed with awe, wonder, respect, and admiration at the power and grandeur of this incredible achievement; the sublime springing from the commonplace.

The principal designer of the bridge, Benjamin Baker, was born in Keyford, Somerset, south-west England in 1840. After early training in a south Wales ironworks, he joined Sir John Fowler in London as a consulting engineer at the age of 21. His early experience was formative. He helped with the construction of the Metropolitan railway in London and he designed the cylindrical vessel in which Cleopatra's Needle, now standing on the Thames Embankment, London, was brought over from Egypt in 1877–8. As his career progressed, Baker became increasingly well-known as an authority on bridge construction. Baker was involved with many of the other great engineering achievements of his day, including the Aswan dam and the Hudson River Tunnel in New York. He was knighted and elected as Fellow of the Royal Society in 1890 and died at Pangbourne, Berkshire, on 19 May 1907.

Sir Benjamin Baker understood the need to explain how his bridge worked, and my account is partly based on his. The three chapters of the Forth Railway Bridge are the towers and cantilevers, the suspended spans, and the foundations. When you first look at the bridge your eyes will probably be drawn to the three enormous towers. In fact they are 104 m tall with large arms or cantilevers stretching out to their sides a distance of 207 m. What you'll probably not notice are the four concrete and stone piers which make up the foundations. Next you might look at the two central spans where the arms of the towers reach out to each other but don't quite meet, and you find a smaller lattice truss structure sitting on each arm. These are the suspended span chapters of the bridge. At each end the arms reach out to the side spans and hence to the river banks.

You'll also probably observe that each tower is a complex latticework with some very large steel tubes and some other much more open and lighter members. As you might expect, every tube and every member has a precise job to do. However, for the moment we want to just keep to an overview of how the main elements of the bridge work.

The outstretched arms are cantilevers just like diving boards at a swimming pool. You can imagine what it feels like to be one of those towers if you hold out your arms horizontally in front of you and hold a bag of shopping or a weight of some kind. You will know instantly that this is not comfortable. There are at least two reasons. First, your muscles are being pulled strongly as you attempt to keep the bag of shopping in the air. Second, the weight makes you feel as though you are tipping forward and you may have to adjust your feet to brace yourself from falling forwards. There is a definite limit to the weight you can hold in this way, which will depend on how big and how strong you are.

If you hold your arms down at an angle to the horizontal you can carry a bigger weight because more of the pull is along the length of your arm. However, that doesn't stop your arms being pushed downwards and rotating about your shoulders. So Baker and Fowler rather cleverly inserted a prop. This prop, however, isn't vertical but rather positioned at an angle to take that downward thrust. Imagine a stiff pole, like a long-handled sweeping brush, under your hand and angled so that the other end is ledged against your foot. If you try this you'll find you can carry much more weight in your shopping bag.

As you add more weight to the shopping bag there comes a time when the weight tends to topple you over. You now need someone to hold your shoulders and prevent you from falling forwards. This is just where Baker and Fowler were again very clever. To illustrate their solution, instead of holding your arms out in front of you now hold them out to the side. With the shopping bag on one side only you will still tend to topple. However, if you put another equal weight of shopping on the other hand, you get a balance and you don't topple. The two weights neutralize the toppling effect. Baker and Fowler calculated that the outstretched cantilevers of the Forth Railway Bridge could carry a small central span truss of 106.7 m.

So in summary at chapter level we can begin to read the flow of forces in the bridge like this. The top arms of the trusses act at an angle to take a pull and the lowers arms of the truss carry the thrust to the foot of the cantilever tower, and the two sides of the towers balance each other. The downward forces from the towers sit directly on and compress the foundations. This idea of cantilever and suspended span is widely used even on much more modest bridges. Around the world many bridges over major highways are reinforced concrete beams with central suspended spans. You can spot them by looking for the bearings at about the quarter points of the central span.

The Forth Railway Bridge is sublime, a magnificent piece of public art. In Chapter 4 we will deepen our appreciation of how it works and how it was built. In the meantime contemplating its beauty as it strides across the Firth of Forth we cannot help but be lifted spiritually from the commonplace. The Sistine Chapel in Rome has the same effect—that is a work of the highest aesthetic, spiritual, and moral purpose. It seems to have little in common with the bridge. Yet bridge building had a direct impact on the composition of the painting on the ceiling of the Sistine Chapel— even that sublime work of art springs from the commonplace. The story of how the Sistine Chapel was painted reveals much about the history of our attitudes to art, architecture, and engineering.

If you have been to Rome to visit the chapel, or even if you have only seen pictures, you must be in awe at the wonderful frescoes on the vaulted ceiling painted 20 m in the air. Furthermore when you hear that it took Michelangelo only four years, you will wonder just how he did it.

Historians of art do exactly that—they try to work out how he did it based on what evidence they can find. Of course there are many different issues they need to consider. How did he get access to the ceiling? Did he plan it all out before starting the painting? What sorts of paint did he use? Did he have any help? Was the chapel used during the four years he was painting?

But since this isn't a book about church art let's just focus on the first question of how he accessed the ceiling. As we will see in a moment many experts say that it involved building bridges and this had a direct impact on the paintings themselves.

Michelangelo was really a sculptor—he had finished the famous statue of David in 1504 (now in the Uffizi gallery in Florence) when, according to Ross King,[9] he was asked—indeed commanded—by Pope Julius III to come to Rome in 1508.

The architect Donato d'Angelo Lazzari, better known as Bramante and at the time hailed as the greatest architect since Filippo Brunelleschi, had devised a system to do the job. Wooden platforms were to be suspended from ropes anchored in the vaults above—but this needed a series of holes to be pierced in

FIG 5. Michelangelo sketch of bridge scaffold for painting of the Sistine Chapel

the ceiling and Michelangelo was not impressed with that idea. He protested to the Pope that this could not work—so the Pope told him to get on and do it himself.

It seems that Michelangelo had had aspirations to be a bridge builder—he had designed a bridge over the Bosphorus in 1506 and for the Rialto Bridge in Venice but neither were built (the present Rialto Bridge was designed by Antonio da Ponte and built in 1591). According to Ross King the now famous sketch by Michelangelo (Figure 5) showing an arch spanning across the chapel is enough evidence to prove how he accessed the ceiling.

But others disagree. The modern dispute started in 1980 when Creighton Gilbert proposed that Michelangelo used a mobile tower. However, such an elaborate system would have closed the chapel which Pope Julius wanted to be kept open.

Indeed King reports on disputes between the clergy and the workmen as both tried to carry on their daily tasks.

In the 1980s Fabrizio Mancinelli and separately Frederick Hartt transformed our understanding of how Michelangelo accomplished his fresco. They presented strong evidence to show that the scaffold was supported by a bridge structure at the level of the clerestory cornice. Mancinelli said there were 12 trusses resting on short brackets projecting from holes in the wall. But this arrangement has a big disadvantage. There was such a 'forest of wood' that progress could not be easily seen from below. The Pope was a regular visitor and he became annoyed that he could see only narrow bands of the vault. So part of the scaffolding was taken down as soon as it was practical to do so.

The consequences of this simple act were profound. Michelangelo realized that some of the figures were too small. So after a lot of thought he made them bigger in subsequent bays. John Beldon Scott wrote, 'Had the scaffold not been removed, the increased monumentality of the later narrative scenes might not have been deemed necessary and we would perhaps have inherited a less heroic depiction of the Creation than the one we see today.'[10] Hartt agreed and said that the bridge had participated quite fundamentally in the artistic process.

Even knowing that Michelangelo worked from a series of bridges spanning across the chapel, how could one man accomplish this enormous task in four years? The answer is that of course he didn't—it was a team effort. There were times when he worked alone but in effect Michelangelo led a team of artisans who worked on every stage of the process. It is misleading to attribute the painting of the Sistine Chapel only to one man. Michelangelo was the inspiration and leader but he could not

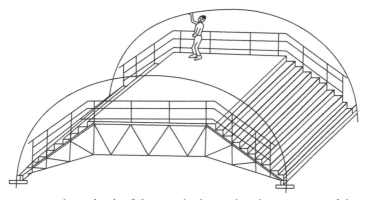

FIG 6. Author's sketch of the truss bridge used in the restoration of the Sistine Chapel from 1981 to 1994

have done it without a team of extremely able, diligent, and brave workers. If you can imagine what it would have been like to climb onto a timber bridge suspended 20 m in the air to work on the ceiling above, you may be able to get a feel for what they had to do almost every day. More than one worker fell to his death helping to create the many Italian frescos we enjoy today.

Figure 6 is a sketch of the truss bridge used in the restoration of the Sistine Chapel from 1981 to 1994. The lightweight aluminium structure was covered in canvas and accessed by lift. It ran on wheels on a track supported by brackets driven into the very holes used by Michelangelo 469 years before.

Most of us, I suspect, think of paintings as autonomous art objects, and painters as lone creators. It certainly came as something of a shock to me to learn that the typical Renaissance painter was often a member of a team that decorated altarpieces, council chambers, townhouses, and palaces. Indeed the Renaissance norm was cooperative production from workshops that

33

fulfilled specific contracts for decorating churches, civic buildings, banners, wedding chests, and furniture. Such contracts often specified not only the size of the painting or statue, the price and date of delivery, but also the subject matter and materials.

So at the time when Michelangelo was working on the Sistine Chapel, it was craftsmen that made the objects we now call art. Indeed our word masterpiece comes from this medieval tradition of an object made by a craftsman at the end of his training to show he had the skills to be called a master. The fresco painters were considered artisans. Contrast this with the modern idea of Michelangelo as a 'fine artist' and genius. In his lifetime there was no such distinction—the concept of fine art came much later.

The story of the development of the idea of fine art is a long and complex one but the separation of art from craft happened relatively quickly. By the end of the eighteenth century the fine artist was almost totally separated from the artisan. All of the 'poetic' attributes such as inspiration, imagination, freedom, and genius were ascribed to the artist and not the artisan. The artisan was simply called a worker in mechanical art—a man with a trade. The modern idea of the masterpiece was completely absorbed into the idea of the artist as a creator—artist-geniuses and their masterpieces. At the same time, science was seen as merely the discovery of how the world works—in other words, not at all creative.

Before about 1850 there was hardly any distinction between architecture and engineering. As more and more science and technical knowledge began to be available people began to take separate roles. Architects took the overview and conceived the overall aesthetic appearance and layout of a building. The growing number of professional engineers looked after delivering the function of buildings—particularly in making it stand up as a structure and in heating, lighting, and ventilation. Engineers were seen by many to be merely applied scientists—doing work that required little creativity. Architecture became associated with the genius of fine art and engineering with the 'mere trade' of the artisans.

Architects and sculptors can and do contribute to the aesthetics of bridges to improve their impact in our public spaces. However, it is misleading to ascribe the design of a bridge to an architect. So when the design of the London Millennium Bridge is attributed in the media to Lord Foster, many engineers will rightly feel that they are not getting proper credit. Bridge building is a team effort largely carried out by engineers. It requires creative thinking to solve difficult problems to make the structure safe as we will explore in Chapter 6. Chris Wise, a member of the original design team for the London Millennium Bridge who now runs his own company, said in 2008, 'There is absolutely no way we would look for an architect to front our bid [to build a bridge]. That would be like Newton asking Shakespeare to front his scientific manuscripts and worse, the public would think that Shakespeare was responsible for the Three Laws of Motion!'[11]

In the next chapter we will look at one of the oldest and naturally beautiful forms of traditional bridge—the arch, the symbol of solidity and security.

2

UNDERNEATH THE ARCHES

Bridges Need Good Foundations

A beautiful bridge is a natural subject for art—the curved shape of the arch has universal appeal. Giotto, said by many to be the father of European painting, used arches to frame parts of scenes in his paintings 700 years ago. One example is St Francis preaching before Pope Honorius III in the church of San Francesco, Assisi, in Italy.[1] Another, the *Dream of (Pope) Innocent III*, shows the church in what appears to be an earthquake—still a modern-day threat, as the building was damaged by an earthquake in 1997.[2] Fra Angelico, painting in 1437, used the subtleties of light and shade underneath the arches for *The Annunciation* in San Marco, Florence.[3]

Until the eighteenth century the story of the arch bridge is the story of the stone or masonry arch (Figure 7). Then bridge builders brought together new materials and new ideas to express the beauty of arches in new structural forms. Two contrasting examples which we will look at later in the chapter are the Salginatobel Bridge in Switzerland (Figure 13), built in 1930, and the 'Squinty' Clyde Arc Finnieston Bridge in Glasgow (Figure 14), built in 2006.

As well as being undeniably beautiful, stone arches are symbols of solidity, stability, and constancy. The countless number

FIG 7. Masonry Arch Bridge at Bradford on Avon near Bath, UK

of them built around the world brings a reassuring sense of timeless unwavering firmness and strength to communities in which everything else seems to be continually changing.

To continue with the book analogy, stone arch bridges are quite easy to read as there seem to be only three chapters. However, there is an important and hidden fourth chapter without which they wouldn't work at all. The first chapter contains the voussoirs, which are the wedge-shaped pieces of stone or brick that form the line of the curved arch known as the arch rib or ring. The second is the roughly triangular area between the outer curve of the arch rib and the bridge deck known as the spandrel. The third contains the foundations, which are always important, but particularly so for masonry

FIG 8. Centring for the Salginatobel Bridge, Switzerland

arches. The unseen hidden fourth chapter is the supporting structure, often called 'centring' that holds all of the components in position while the arch is being built. Figure 8 shows the centring scaffolding built by only six men in 1929 to support the arch of the Salginatobel Bridge (Figure 13) as the concrete was being cast. It isn't until every piece of the bridge is in place that the props can be removed and the flow of internal forces can begin to work. The arch derives its strength from its shape; it will stay in place for a very long time as long as the foundations don't move. That is the underlying reason why stone arches are robust and long lasting.

The first stone arch bridge in London was completed in 1209 after 33 years of construction work. Probably the first London Bridge was built in timber by the Romans around AD 60, although bridge historian Frederick Robins says the first definite evidence of a bridge is from as late as around AD 1000.[4] After the Romans left, the timber bridge was repaired several times. King Olaf of Norway demolished it in about 1014 in an attempt to help King Aethelred regain London from the Danes. Subsequent bridges were destroyed by storm in 1091 and fire in 1136. The words of the nursery rhyme 'London Bridge is falling down, my fair lady' was probably inspired by the number of replacement bridges that had to be built. Early bridge builders feared that their intrusions into the river might upset the river gods and so often made sacrifices to appease them. It is possible that the words 'my fair lady' refer to one way in which the bridges were 'magically' strengthened by burying a dead virgin in the foundations of the bridge. Robins states that the idea of a 'foundation sacrifice' to appease the river gods was widespread.[5] He reports that as late as the nineteenth century, Balkan folklore required that a dead child be buried under a bridge at Trebinje.

King John allowed houses, shops, and a chapel to be built on the first stone arch of London Bridge. Contemporary pictures show them up to seven stories high. The bridge consisted of 20 small arches with a drawbridge and gatehouse at the southern bank. The water flow was so restricted by the arches that the 'rapids' formed under the bridge were dangerous. It was said that the bridge was 'for wise men to pass over, and for fools to pass under'. The houses and shops took up so much space on the bridge that crossing it could take an hour if a cart broke down or an animal broke loose. In 1263 Queen Eleanor, who apparently wasn't particularly popular at the time, was travelling by water from the Tower to Windsor, when she was pelted by a mob from the bridge. Perhaps the reason was that the Queen seemed to regard the bridge as a source of revenue rather than a responsibility, and the condition of the bridge deteriorated. As a consequence, when in 1281, the Thames froze over, five of the arches collapsed under the pressure of the ice.

The buildings on the bridge were a fire hazard. Three thousand people were reportedly killed in 1212 after being trapped in the middle of the bridge by fires at both ends. Six tenements were destroyed by fire in 1504. Another big fire broke out in 1633 and the northern third of the bridge was destroyed. However, the gap created by that fire did prevent the Great Fire of London from spreading to the bridge in 1666.

By the end of the eighteenth century it was clear that a new bridge was needed. A competition attracted entries from eminent engineers such as Thomas Telford, but was won by the equally respected John Rennie. It turned out to be his last project—indeed his son John, who was later knighted for his engineering achievements, completed it. The new bridge was built 30 m upstream of the old one and was opened after seven years of building work in 1831. HMS *Beagle*, which later took Charles Darwin on his famous expedition, was the first ship to pass under one of the five new arches.

John Rennie the older was born on the Phantassie estate, near East Linton, 20 miles east of Edinburgh, in 1761. John was only six years old when his father, a prosperous farmer, died. The young budding engineer was fascinated by machines and loved to make working models. At only 12 years old he left school but realized his mistake, went back and then from 1780 to 1783 studied natural philosophy and chemistry at the University of Edinburgh. In 1784 John decided to go to England to find new opportunities. At first he worked for Matthew Boulton and James Watt at the Albion Flour Mills in Blackfriars, London, where he pioneered the use of cast iron. In 1791 he set up his own business as a mechanical engineer in Blackfriars. He worked on the building of several canals, but some of his largest projects were for the Royal Navy which at that time was beginning to build the infrastructure for its century of world domination. He designed three bridges over the Thames, all now replaced: Waterloo Bridge, which had nine equal arches and a flat roadway; Southwark Bridge, which was the widest cast-iron span ever built in Britain; and, of course, London Bridge. His eminence was such that, by 1798, he was elected a Fellow of the Royal Society. He died in 1821 at the age of 60 and was buried in St Paul's Cathedral.

Fortunately the nursery rhyme doesn't refer to Rennie's London Bridge or the nearby Tower Bridge with which it is sometimes confused. Indeed when Rennie's Bridge was famously sold to Robert P. McCulloch in 1968 and reassembled in Arizona it was, perhaps mischievously, rumoured that Mr McCulloch thought he was buying Tower Bridge. The present London Bridge, opened in 1972, is not an arch bridge but a beam bridge—so we will leave the story there until we consider beams in Chapter 3.

The Ponte Vecchio—the *Old Bridge* in Italian—is another historically important medieval stone arch bridge over the Arno River, Florence, Italy. Like the old London Bridge it is a natural rendezvous and place to shop—especially for jewellery. The original wooden Roman bridge was destroyed by a flood in 1333. A stone replacement, designed by Taddeo Gaddi, was constructed in 1345. It has three arches—spanning 30 m in the centre and 27 m at the two sides. It seems that the jewellers that set themselves up on the bridge were exempt from a form of tax. They were allowed by the Bargello—the Florentine equivalent of a chief of police—to show their goods on tables. The idea of bankruptcy is said to have originated from this practice. When a shopkeeper could not pay his debts, 'the table on which he sold his wares (*banco*) was physically broken (*rotto*) by soldiers, and this practice was called *bancorotto*.' A *banca rotta* is a broken counter or table of a bankrupt merchant.[6] Like all shopkeepers those on the bridge sought new ways to promote their goods. The owner of a padlock shop at the end of the bridge encouraged lovers to fasten a padlock onto the bridge and to throw the key into the river thus guaranteeing an eternal bond between them. As a consequence thousands of padlocks frequently had to be removed and the bridge structure was damaged but a small fine soon reduced the problem. When the Germans retreated from Italy in 1944 during World War II, Hitler ordered that the Ponte Vecchio should not be destroyed and so it remains as one of the many symbols of the magnificent city of Florence.

The Romans were the first prodigious stone arch bridge builders. One of their most famous extant structures is the multi-arch Pont du Gard across the Gardon Valley in Provence, built by Agrippa in the middle of the first century AD to carry water to the city of Nemausus (Nîmes). They used no mortar or other binding agents except a few iron clamps. So how did they do it? How can such an impressive structure work without any mortar at all?

The secret of the stability of stone or masonry arches lies not in the mortar but in how the stones are packed together to allow the internal forces to flow. Dry stone walls are so-called because they have no mortar. So let us now start to read masonry arches by first considering how dry stone walls stand up.

Dry stone walling is a well-known feature of the English countryside especially in the Derbyshire and Yorkshire dales and in the Cotswolds. The builders dig a shallow, narrow trench and lay a base of small stones. The wall is then built up in horizontal layers, each narrowing slightly towards the centre of the wall. The centre is filled with small stones or rubble, and part way up stones are laid across the width of the wall to tie it together. The topping is generally a row of slanting or vertical stones.

If you try to push over a dry stone wall you are testing one of the structural principles of an arch. The reason both the wall and a masonry arch are stable is simply that they are heavy and wide. Although this is fairly obvious for a dry stone wall and a stone arch bridge it is perhaps less clear for the slender and towering arches of a cathedral—but we will come to that later.

Each stone in the wall bears on the stones below and hence compresses them. This pressing down or compressive force gets bigger as you go down the wall and follows what is called a 'line of thrust', which is the line along which the internal forces flow. At the very top there is no force so the thrust line effectively starts through the thickness of the top stone. At any point in the wall the thrust line is due to the weights of all of the stones above and is vertical and downwards and acts at the centre of gravity— the point which represents the weight of all of the stones at a given level in the wall.

When you push the wall sideways the thrust line becomes the combined effect of your push and the weight of the wall. As a result the thrust line is inclined at a very small angle to the vertical because normally your push will be small compared to the weight of the wall. In that case the thrust line will not be seriously compromised and the wall will remain stable. In order to push the wall over completely and as a whole, you would have to disturb the thrust line so much that it no longer passes through the base of the wall. In other words you would have to push the wall past its tipping point. What this means is that you will have pushed it so far that the centre of gravity of the entire wall has moved over and past the edge of the base of the wall. In fact this is almost impossible. Before you get to the tipping point the stones you are pushing against would simply slip against each other so that each stone would move and fall separately depending on where and how you actually pushed the wall. You would cause a localized fall of stones rather than a tipping of the entire wall. We'll see that effect in detail for a corbelled wall in a moment.

So what have dry stone walls got to do with stone arches? Nowadays if we want a gateway through a brick wall we would put a timber or steel lintel beam over the opening and build the bricks over it. The ancients, however, used a technique called corbelling. Dome-shaped tombs were created this way. First the builders laid out some stones or bricks in a closed shape such as a circle. Then a second layer of stones was built on the first layer with each inner stone projecting inwards just a little. This was repeated for successive layers so that, as they built higher and higher, the layers became closer and closer and the enclosed area became smaller and smaller. Finally when the layers nearly met at the centre they capped the hole with a larger stone. As you might imagine the technique had severe limits since each stone

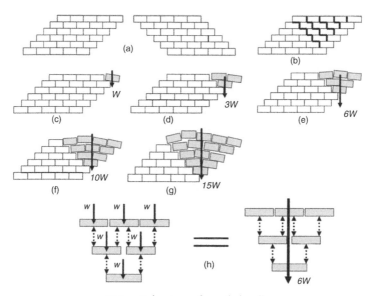

FIG 9. Behaviour of a corbel wall

cannot project out very far otherwise it would tip over and fall inwards. Figure 9a shows a side view cross section of six layers. I have drawn regular bricks rather than irregular stones just to make it easier to read. In the diagram the two sides do not quite meet at the top and I haven't shown the capping slab.

So how might such a wall tip over and collapse? In passing, it is worth noting that this is an important question for bridge builders. They are constantly asking themselves 'How might my bridge collapse?' If they know what could happen then they can work out how to stop it.

To keep things simple let's think about how a corbelled wall might collapse if it is just one brick wide and hence not a dome-shaped tomb but just a simple wall. Figure 9b shows one part of a corbelled wall—the same as (a) but only one side and one brick wide. The diagram also shows the possible fault lines in bold if the bricks project out too far. The first and simplest example of this is shown in Figure 9c. It happens when the top brick tips if its centre of gravity goes beyond the edge of the brick underneath. The brick has gone past its tipping point. Diagrams (d) to (h) in Figure 9 show the same tipping effect but for the other fault lines with more and more bricks involved each time. All of these mechanisms are possible ways in which the wall could fail. In each case the bricks fall away if the effective weight of all of them acting together falls outside the edge of the bottom brick—the tipping point. We'll soon see how important this tipping point is for arches too.

We know that reading a bridge requires us to think about how its internal forces flow through it. Figure 9h shows the external and internal forces between the six bricks of Figure 9e when they fall together as a group. Remember that each brick exerts a downward force of its weight, W. The top three bricks push down on the middle two which push down on the lower one. The internal forces are compressive since the lower bricks have to push back against the upper ones and they are shown as dotted arrows. The total downward thrust of all of the bricks together is $6W$ and acts through the centre of gravity of the group—right in the middle. If this line of thrust falls outside of the edge of the supporting brick then the group will tip as a whole, as shown in (e).

It is important to understand that corbelled structures are not arches but they were the forerunners of arch construction. We don't know who actually first developed the arch although it is known that the Etruscans used them before the Romans. The conceptual leap from corbelling to arches is actually quite considerable because, as I said earlier, arches must be supported by another structure whilst they are being built. This supporting structure can be of many forms—usually it is a timber truss (see Chapter 4). The details don't matter as long as it holds the voussoirs until they are all in place. Usually the stone in the centre is the final stone to be set and so it is called the keystone and it is often bigger than the rest. When the keystone and all of the other voussoirs are in place then the centring can be removed.

Naturally the voussoirs must be shaped to fit together to form the curve—they are not rectangular. Each stone interacts with its neighbours just like the dry stone wall but the thrust line is more complex. Each stone is pushed down by the stone above and can only push back if it can push down on the stone below. That is the reason why arches only work when they are complete: each voussoir reacts off its lower neighbour. The very bottom stone must react against a foundation, so it must be firm. The whole story is a simple example of 'togetherness': stones working together as a group to allow the internal forces to flow in balance. Individual stones on their own don't make an arch but arrange a group of stones in a certain configuration and they can work together to make an arch.

The difference between the arch and a dry stone wall is that in an arch the direction of the line of thrust changes as the force from the stone above combines with the weight of the voussoir plus the weight of any infill rubble or stones from the spandrel. These three forces combine to bear down on the stone below and the stone below pushes back. By contrast the thrust line of a wall, corbelled or not, is vertical through the thickness of the bricks.

So how do we know where the thrust line goes in an arch? Robert Hooke found the solution in 1675. He realized that if you hang a chain between two points then you get an upside down version of an arch. He wrote, 'As hangs the flexible chain, so—but inverted—will stand the rigid arch.'[7]

Imagine a single length of rope or cable, or even a bicycle chain, hanging from two points with no weights attached. The shape it makes is the 'line of pull', called a *catenary*, and results only from the self-weight of the chain. Each link of the bicycle chain or each little element of the rope or cable is being pulled by the links or elements on either side—its neighbours—in a way that balances its own weight. Now imagine that chain upside down. The shape it forms is the exact opposite of the line of pull—it's a line of thrust. If we now hang the chain as before but load it with weights equal to each of the voussoir stones at points along the chain then we get a new 'line of pull' shape. That new shape inverted is the line of thrust of the arch because it defines the line of action of the force of each voussoir stone as it pulls/pushes on its neighbours. Just as a chain is in tension, so an arch is in compression. Arches are upside-down squashed chains.

Just as the line of thrust for our dry stone wall had to lie within the base of the wall for it to remain stable, so the thrust line for the arch must fit within the line of the voussoirs of the arch. We'll see what happens if it doesn't in a moment.

Before we do, we need a bit more bridge grammar. Scientists distinguish force and mass. They describe a force as a vector and mass as a scalar. This seemingly unnecessary distinction turns out to be incredibly useful in calculating the thrust line of an arch.

A vector is anything that has size and direction. So, for example, your velocity in a car is a vector because it has a size (say 30 mph) and a direction (say due north). A scalar has only size. The speed of your car is normally a scalar because it is not associated with direction. Acceleration is the rate of change of velocity (vector) or rate of change of speed (scalar). Velocity is measured in metres per second (m/sec), so acceleration is so many metres per second per second or m/sec^2.

A 1 kg mass of a bag of potatoes is a scalar. However, the weight of that bag acts downwards and so is a force vector. Sir Isaac Newton's second law of motion tells us that a force is a mass times an acceleration. So the potatoes, just like any other object on the Earth's surface, if free to fall to the Earth, will do so with an acceleration of 9.81 m/sec^2. Therefore the weight of our bag is 1 kg times 9.81 m/sec^2. A newton (N) is defined as 1 kg m/sec^2 so our 1-kg bag of potatoes weighs 9.81 N.

We can represent a vector on a diagram using an arrow. For the car example you simply draw the arrow in the same direction (say north) with a length proportional to the size (say, 5 mm = 10 mph). Your arrow for 30 mph will therefore be 15 mm long. In the same way a horizontal force of 10 kN can be represented by a horizontal arrow say 10 mm long. The scale in that case is therefore 1 cm = 10 kN.

Newton's third law of motion says that action and reaction are equal and opposite. If you push me with a certain force then I must be pushing back with an equal and opposite force. So, in an arch, each of the voussoir stones is pushing against its neighbours, and they are pushing back on it.

Each voussoir stone has four forces acting on it: the weight of the stone, the weight bearing down on the stone from the mass of masonry above it, and the reactions from the two adjacent voussoir stones. However, to simplify things, we can combine

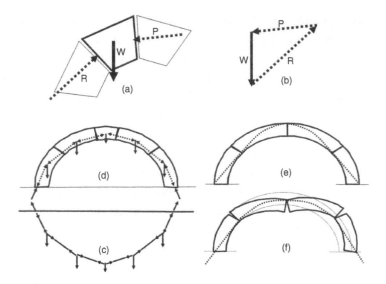

FIG 10. The forces in a masonry arch

the weight of the masonry plus the stone's own weight into a single force W. This is shown in Figure 10a, which also shows the reaction force P from the neighbouring voussoir stone pressing down on it, and the reaction force R from the voussoir stone below. If equilibrium is to be maintained, these three forces must balance each other out.

Of course, in turn, our stone will bear down on the voussoir immediately below it with an equal and opposite force R. In this way, the forces between each stone and the next define the thrust line in that part of the arch.

We can draw arrows of the three forces to make a closed loop— a triangle of forces as in Figure 10b. It is drawn so that the directions of the forces follow each other around the triangle. If the loop closes then the forces balance out and are in equilibrium.[8]

Figure 10d shows that all of the voussoirs in an arch bridge have these three forces acting on them. Figure 10c is Robert Hooke's hanging chain. The thrust line in the arch in (d) is simply the upside-down version of the tensions in the chain (c). Both define the flow of internal forces. Note that the external forces at the base of the arch (shown as full arrows) are the reaction forces from the foundation and must balance the total weight of the arch. If they don't the arch will slide and may fail.

There are many possible thrust lines within the arch. Fortunately we don't need to know exactly where the real one is as long as we know it lies within the voussoirs from the keystone at the top to the springing stones at the base. Figure 10e shows the arch at the limit when the thrust line is at the very edge of the voussoirs in five places. When the thrust line force is at the very edge of the stone then it is at the tipping point. If the thrust line goes outside of the line of the voussoirs, as in Figure 10f, then the voussoirs will rotate about the edge—rather like a door hinge.

Once an arch bridge is built and in place then we know that the thrust line is within the voussoirs—that is our starting point. The key to the continued success of the bridge is then in the foundations. The thrust at the base of the arch will be large— approximately half of the total weight of the entire bridge. As long as the reactions from the foundations are such that the thrust line is near the middle of the voussoirs then everything will be fine. However, if the foundations cannot cope and move slightly outwards then the thrust line will also move. If it strays to the edge of the stones then the joint at that point will open up and a crack will appear. If only one hinge forms in the arch then it may be a bit unsightly, but the arch will not collapse. Indeed if three hinges form, the bridge will not collapse either.[9] For the whole arch to collapse the arch must have four or more hinges because then it becomes a mechanism—an assembly of moving parts as shown in Figure 10f. Normally the foundation movements and consequent cracks and hinges in the masonry do not happen quickly so there is time for those people in charge of the bridge to read the signs and make repairs.

The actual forces on each small area of a stone within the arch may be quite small. So big arches with big stones may have small stresses but carry very large weights. As you can see from the triangle of forces in Figure 10b the weight is a crucial part of a masonry arch because it strongly influences the position and direction of the thrust line. If W is large then the direction of R will tend to be more vertical. Medieval church builders used this idea by building pinnacles on top of arches to give extra weight just where it was needed. For example, you commonly see pinnacles on top of a flying buttress to 'turn down', or make the direction of the thrust line more vertical and so keep it within the line of the arch or vault (vaults are arches in two directions). Church buildings like Westminster Abbey are clever pieces of medieval structural engineering because of the way the positions of the various thrust lines are controlled.

Of course the medieval builders didn't have the benefit of our modern understanding of the triangle of forces so that the techniques they used to control thrust lines were learned by long experience of trial and error. When these hard-earned rules were ignored then there could be severe consequences. The story of Beauvais Cathedral illustrates this nicely. Of course a cathedral is not a bridge in the normal sense of the word. Nevertheless the roof is a towering masonry arch which bridges the interior of the church. The whole cathedral itself is a symbolic bridge between man and God.

The first bishop of Beauvais was Saint Lucien in the third century AD. After he died, a number of churches were erected over his tomb. Eventually ambitious plans were drawn up to build a massive new Gothic cathedral, a political act of defiance to assert independence from the French king. The building of the apse and choir had already started in 1247 when a new bishop, William of Grez, arrived. He had even greater ambitions and decided that he wanted the tallest cathedral in Europe. So the medieval masons added an extra 4.9 m to the height. But just as work began to fulfil the bishop's wish, in 1248, the choir vault came crashing down.

It seems that from the first, before Bishop William arrived, the building work had struggled. Workers came and went, apparently falling out of favour. Money was tight and priorities kept changing—classic conditions for a project to go wrong. The irony was that the arrival of Bishop William had reinvigorated the project. Whatever the actual reason for the fall, it was certainly believed at the time that the spacing between the piers or pillars was too large. New intermediate piers were built between the originals so that the bays were halved from a spacing of about 9 to about 4.5 m.

The work was interrupted by the Hundred Years War and by the English occupation so a start was not made on the transept until 1500. When it was well under way, in 1544, the Bishop and Chapter (the church assembly) decided that they needed a tower over the crossing—at the intersection of the nave and transept. Experts were called in to help decide whether to build it in stone or timber. Models were made and talked over. In 1558 they decided on a masonry tower. Construction work began in 1564 and finished in 1569. The new tower was big—153 m tall. It alarmed many people so the Chapter commissioned some examinations. After two years a detailed report by two of the King's masons said that the four main crossing piers were beginning to lean. They said that two nave bays were needed and that the foundations should be strengthened and temporary walls built between the crossing and the piers.

Unfortunately the Chapter dithered, took further advice, and only approved the work after a delay of two years. Thirteen days after the decision, on Ascension Day, 30 April 1573 the tower fell. Fortunately no one was killed even though a procession had just left the cathedral. The Chapter decided that was something to celebrate so they gave thanks to God for the safety of the faithful of Beauvais. In 1605 they decided to live with the existing structure and Beauvais became a choir and transept without a nave—as it is today.

The tower was probably never really in equilibrium. As the structure slowly drifted it was restrained by tensile and shearing stresses in the mortar and some interlocking stones. Eventually, however, the drift became so large and the columns pushed so far out of true that the inevitable happened.

The stress levels in most of an arch structure are so low that the strength of the material at Beauvais was not the main concern. But the shape of the arches was critical because that is the way the thrust line, the flow of internal forces, should have been controlled. Of course this was before Newton had formulated his laws. At that time the importance of shape was expressed by numerical rules of proportion that had been developed over a long period and found by hard experience of trial and error. At Beauvais this experience was lost or ignored for aesthetic reasons. Once this had been done there was no way of reconstructing the rules by any process of thought or intellectual argument.

Designers of arch bridges have a heavy responsibility to make their bridges safe and the story of Beauvais Cathedral shows that they must not bend the rules too far. However, when new materials and new ideas become available, they create situations in which there aren't any rules to go by. Traditional masonry arches are very heavy and hence costly and time-consuming to build, and maximum spans are limited. So it is unsurprising that when cast iron became available in sufficient quantities, new forms of arch were suggested. Wood was replaced by coke for the smelting of iron by Abraham Darby from about 1709 onwards. His grandson Abraham Darby III built one of the first non-masonry arch bridges, over the River Severn in Shropshire at Ironbridge in 1777–9 (Figure 11). As its name implies it was made in cast iron but it followed the traditional form of a masonry arch because that was the natural form of a bridge at the time.

The only way to cross the Severn Gorge in the early eighteenth century was by ferry. Coal mining, iron foundries, and earthenware manufacture in Coalbrookdale and Broseley on opposite sides of the river had to be better connected so it was entirely natural to think in terms of cast iron for a bridge. Thomas Pritchard suggested and designed the bridge and in 1773 Abraham Darby III was given a commission to cast and build it. Unfortunately for him, Darby agreed to fund any overspend. The project turned out to be far more expensive than estimated and so Darby was in debt for the rest of his life. Nevertheless his memory lives on in his magnificent bridge which was opened on New Year's Day 1781.

FIG 11. Ironbridge at Coalbrookdale, UK

Very large castings were needed for a structure spanning 30.5 m (100 ft) and rising 18.3 m (60 ft) above the river. Looking at this bridge (Figure 11), the first thing you notice is that the voussoir stones have been replaced by three cast-iron ribs in five rows across the bridge. They appear to be linked almost as though the spaces are the voussoir stones themselves. The ribs carry a compressive force just as the stones but in a different way. We'll need to come back to just how they do this when we consider truss structures in Chapter 4. Each half-rib was cast in one piece weighing around 6 tons and then the two halves connected at the top in the middle. The bridge deck is carried down onto the arch by other cast iron pieces, one vertical, one circular, and one shaped like an upside-down V. The whole bridge is supported by a massive abutment to resist the horizontal outward thrust. In fact over 800 castings of twelve basic types were used. Each member was cast separately and the fastenings were of the form used by carpenters. In other words the new bridge was made of a new material but built in an old and traditional form.

Unfortunately after just a few years the ground had moved and cracks appeared in the masonry abutments. In 1802, the southern abutment was demolished and replaced with temporary wooden arches which were eventually replaced by iron arches. Major repairs on the foundations were carried out in 1972 and again in 1999–2002.

Cast iron is brittle and therefore limited as a structural material. In 1784 Henry Cort produced wrought iron in a coal-fired flame furnace through a so-called 'puddling process' in which the carbon and other impurities that makes the raw 'pig' iron brittle were burned off to make the more malleable wrought iron.

Isambard Kingdom Brunel understood the potential of wrought iron. He was the engineer for the Great Western Railway and for other allied companies such as the Cornwall Railway and thus was responsible for a number of bridges. Harold Hopkins wrote that three in particular—Windsor, Chepstow, and Saltash—could be regarded almost literally as Brunel thinking aloud as the spans got larger.[10] At Windsor gone was the heavy infill of the old stone-arch spandrel. Instead Brunel designed a truss latticework (Chapter 4) suspended from an arch with the deck hanging beneath. He cleverly used the railway deck to resist the outward thrust, normally taken by the foundations. In effect the deck became a large tension member holding the two ends of the bridge together. He took this a stage further at Chepstow where he used suspension links not only to resist the outward thrust of the arch but also to help in carrying vertical loads. So he

FIG 12. Brunel's Saltash Bridge

designed a truss bridge with a top chord forming a flat arch and
three tension links to form an apparently continuous chain
hanging from the top of the end posts of the truss. Then in
1855 he designed his final masterpiece, the Royal Albert Bridge,
or Saltash Bridge (Figure 12), to carry trains over the River Tamar
between Plymouth, on the Devon bank, and Saltash on the
Cornish side. This combination of an arch with suspension
chains had clearly developed out of the previous two. It has
two 139-m main spans 30 m above mean high spring tide with
seventeen shorter approach spans. It was completed the year
Brunel died and is still in active service. Onlookers are left in
no doubt who designed the bridge as the letters 'I K BRUNEL

ENGINEER 1859' are clearly there for all to see on the end of the structure.

When you look at the bridge for the first time your eyes are immediately drawn to the two massive arch tubes that span from one abutment to the central pillar—the first chapter. The tubes are a far cry from stone voussoirs but they carry compressive forces similar to those we discussed earlier, but now as gigantic beams. However, we'll have to wait until Chapter 3, where we look at how beams work, before discussing them in detail. The arch tubes were constructed from half- to three-quarter-inch wrought iron plates riveted together. If you were to slice one of the tubes vertically you would find an oval-shaped cross section 5.1 m (16 ft 9 in.) wide and 3.7 m (12 ft 3 in.) deep with a diaphragm to provide some stability at each end. Brunel knew that this was efficient to take the large compressive forces and to prevent the arch from bowing out sideways. The aerodynamic shape also helps to reduce the load from the winds that blow up the Tamar Valley.

After the tubes there are two suspension chains on either side of the arch—the second chapter. The chains and arch tubes meet at the abutments and at the pier. Brunel is cleverly combining the principles of a suspension bridge, which we'll look at in Chapter 5, with a beam arch. The suspension chains are a series of wrought iron links. The end of each link is enlarged to form an eye with a hole. The links are pinned together using a wrought iron pin 100 mm (4 in.) in diameter.

The third chapter of the Saltash Bridge contains the vertical suspenders. These effectively replace the stone infill in the spandrel of a stone arch. They are the means by which the internal forces are transmitted or flow from the bridge deck to the arch and to the suspension chains. The bridge deck, on which the trains travel, just hangs from the arch and the chains. It consists of two wrought-iron girders each 2.4 m (8 ft) deep and spanning 141.7 m (465 ft). Between these girders are cross girders 5.1 m (16 ft 9 in.) long and 0.33 m (13 in.) deep. Overall some 2,650 tons of wrought iron were used with 1,200 tons of cast iron and some 17,000 cubic metres of masonry and brickwork.

So how do the internal forces flow? If you imagine a train on the bridge deck then the weight of the train bears down on the bridge deck. This in turn causes a tension in the hangers. The hangers pull down on the arch and on the suspension chain links. The arch resists the pull from all of the hangers and therefore spreads out slightly producing an outward thrust at its ends. The chains resist the pull from the hangers and sag slightly causing an inward tension at its ends. The inward pull on the chains counteracts the outward thrust from the arch. The whole system is to a major extent self-balancing. This use of arches and suspension chains together was an ingenious solution by Brunel but the idea never caught on and few other bridges of this type remain.

The fourth chapter of the bridge contains the abutments and foundations which are, as always, very important. Brunel was particularly innovative in his use of compressed air caissons to build the central pier. The foundations were tricky. The bridge had to be supported 24.4 m (80 feet) below sea level so Brunel used a giant cylinder as an air-tight caisson that was floated out and sunk onto the bed of the river to form the basis for the central pier. The air of the caisson was under sufficient pressure to exclude the water so that the men inside could work in the dry. At that time some of the effects of working under such conditions were known but the basic nature of the problem was not understood and the men suffered from cramps and fatigue. Inside the caisson they excavated soil to bedrock and then built up the masonry pier ready for the bridge.

The two 138.7-m (455-ft) bridge spans were built on the shore, floated into position on pontoons and then jacked up. Massive hydraulic cylinders under the centre of each end of the truss lifted the span 0.92 m (3 ft) in one day—one end at a time. The masonry was progressively built up under the supports after each lift at each end. Cast-iron octagonal columns were used for the supports at the central pier.

While Brunel was building his bridge Henry Bessemer (later Sir Henry) was working on a new idea for making steel. In 1855 he replaced the puddling process with a mechanical process of blowing a blast of air through the fluid pig iron. This eventually led to the development of steel production in large quantities at economic prices. William Kelly independently discovered the process in the USA at about the same time but unfortunately went bankrupt trying to develop it. The first Bessemer steel bridges were built in Holland but unfortunately the steel was of poor quality and there was a prejudice against steel for some years.

Perhaps one of the most iconic steel arch bridges is the massive Sydney Harbour Bridge, Australia. This bridge, known locally as the coat hanger because of its shape, is all the more remarkable for being built at all since the work was done during the Great Depression. It carries eight lanes of road traffic, two railway tracks, one footpath, and a bicycle track over a span of 503 m and so is one of the widest long-span bridges in the world. The first plans for a bridge had been drawn up as early as 1815.

Some building work began in 1890 but it wasn't until 1911 that a formal proposal was accepted from John Bradfield for a cantilever truss bridge. As a result he was appointed chief engineer for both the bridge and Sydney metropolitan railway construction. Unfortunately World War I intervened and so all work was put on hold. Bradfield travelled extensively looking at bridges around the world. In New York he saw the Hell Gate Bridge built in 1916 and immediately recognized new possibilities. In 1922 the design and construction of the Sydney Harbour Bridge was put out to tender. The specification allowed for both an arch and a cantilever bridge and defined all the conditions to be satisfied but it did not include designs. Dorman Long and Co., of Middlesborough, UK, won the job. Their consulting engineer was Sir Ralph Freeman, whose company, Freeman Fox and Partners, later designed the ill-fated Westgate Bridge (see Chapter 3). Work began in 1923 but construction of the arch didn't begin until 1929.

The bridge has four chapters, the arches, the suspenders, the bridge deck, and the foundations. They work together just as the Saltash Bridge but without suspension chains. Instead of large tubes the arch is built of two giant trusses with 24 panels that vary in height from 18 m at the centre to 57 m at the ends. The summit is 134 m above sea level. Interestingly that distance can increase by as much as 180 mm on hot days as the steel expands. Two large metal hinges with pins 368 mm (14.5 in.) in diameter at the base of the bridge carry the massive horizontal thrust of around 20,000 tons from the arch. The hinges are housed in the pair of pylons, or monumental towers, at each end which are about 89 m high and are made of concrete and granite. The pylons are not part of the structure—they were added primarily to add visual balance. All four have now been turned into a museum and tourist centre and a lookout over the harbour. The bridge deck is 1.15 km long—cast in concrete and lying on beams that run along the length of the bridge. These sit on steel beams that span the width of the bridge—you can see them from below.

Two separate construction teams built the two sides of the arch as cantilever trusses. Wire cables anchored back through inclined U-shaped tunnels cut into the rock held the trusses in place. Cranes at the ends of the cantilevers lifted each steel piece into position and the bridge builders riveted them into place. The cranes then 'crept' forward onto the new section and the whole process was repeated. The southern end was worked on a month ahead of the northern end to detect errors and to learn lessons for the northern side. When the cantilevers were completed the two sides were nearly touching. The cables were then slowly released and the two halves gradually brought together at the bottom of the truss and pinned together. At this point the arch had effectively 3 pins—one at each end and one in the middle (see Chapter 3 for the significance of this). The two top chords of the truss were then jacked apart to a predetermined load that produced similar levels of stress in the top and bottom chords and the support cables removed. The bridge then became a two-pinned arch. The last stage was to connect the vertical suspenders to the arch and, starting at the centre, to sling the bridge deck from them.

The road and the two sets of tram and railway tracks were completed in 1931 together with other services such as power and water. In early 1932, the first test train, a steam locomotive, safely crossed the bridge and a series of such tests followed. By today's standards safety for the workers was poor. Sixteen died during construction—mainly by falling. Several were injured during riveting and some became deaf. The bridge was formally opened on 19 March 1932. The whole project was dogged by the political and class conflict of the time brought on by the Great Depression. The Prime Minister of New South Wales, Jack Lang, had been elected on a policy of not paying back the British loans but instead using the money to keep people employed in public works. The federal government denounced this as illegal but was consequently brought down by Lang supporters. At the opening, instead of lambasting the federal government as many expected, Lang was statesmanlike as he referred to 'the people's bridge'.

> The achievement of this bridge is symbolic of the things Australians strive for but have not yet achieved … Just as Sydney has completed this material bridge, which will unite her people, so will Australia ultimately perfect the bridge which it commenced 30 years ago at Federation … the bridge of understanding among the Australian people will yet be built.[11]

Despite these fine words the Governor of New South Wales eventually dismissed Lang's government.

The bridge has inspired many an artist. The Australian poet C. J. Dennis wrote

> It 'appened this way: I 'ad jist come down
> After long years, to look at Sydney town,
> An' 'struth! Was I knocked endways?
> Fair su'prised?
> I never dreamed! That arch that cut
> the skies![12]

It was a long way from the modest stone arches to the magnificence of Sydney Harbour Bridge. Cast iron, wrought iron, and steel in turn had stimulated completely new forms of bridges. At the same time another quite different material, concrete, was having just the same effect. A cement-based mortar had been used for centuries to bind masonry together. Vitruvius, a Roman architect and builder, wrote about pozzolana ash, a rich volcanic deposit found near Naples and Rome, as cement. However, it wasn't until 1796 that James Parker patented a form of cement that he rather misleadingly called 'Roman cement'; it was obtained by burning limestone from near the River Thames. Then Joseph Aspdin made lengthy experiments and eventually succeeded in making the first artificial cement by burning a mixture of clay and lime. He patented it in 1824 and called it Portland cement.

Concrete was a much more plastic material than masonry in the sense it could be more easily moulded into any required shape and so it was quite natural to build an arch bridge using mass concrete instead of masonry. The first example was in 1865 for the multiarch Grand Maitre Aqueduct taking water from the River Vanne to Paris.

However, it was soon realized that there were even better options with this new material. In 1808 Ralph Dodd had proposed embedding wrought iron bars to give the concrete greater strength in tension but Parker's cement was a bit too crude for this to work. By the 1850s a number of patents had been taken out for reinforced concrete. In 1875 Joseph Monier built an arch bridge 16 m (52.5 ft) long at Chazelet Castle in France but there was no theoretical knowledge about how this should work. Thaddeus Hyatt carried out some early tests in the USA and showed clearly that reinforcement should be placed at the bottom of a simply supported beam (see Chapter 3).

As we have seen, one obvious way to reduce significantly the weight of a stone arch bridge was to take out much of the spandrel infill material and create an open spandrel bridge. Indeed the idea had been used much earlier in one of the oldest standing bridges of this type, the Zhaozhou Bridge built in 595–605 in Hebei Province in China. It is a segmental arch—which simply means that the arch is less than a semicircle. The rise is only 7.3 m or 0.2 of the span of 37 m. To compare—the rise of a semi circle would be 18.5 m—it was a very flat arch for its time.

Reinforced concrete began to open new possibilities for open spandrel arch bridges. Francois Hennebique and the German engineer G. A. Wayss were amongst the first designers to exploit them. Hennebique's Vienne River Bridge at Châtellerault, France, built in 1899, was the longest spanning reinforced arch bridge of the nineteenth century. In 1904 the Isar River Bridge at Grüne-wald, Germany, designed by Emil Morsch for Wayss's firm, became the longest reinforced-concrete span in the world at 69 m. The French engineer Eugène Freyssinet designed a number of bridges and went on to develop prestressed concrete as we shall discover in the next chapter.

However, it was perhaps Robert Maillart who made the greatest aesthetic impact. He worked briefly with Hennebique before establishing his own business. He went on to design slender arch bridges such as the spectacular Salginatobel Bridge in Switzerland in 1930 (Figure 13).

The chapters of this bridge are as before except that the voussoirs are replaced by the rib of the arch which is a reinforced concrete beam. The bridge deck is also a beam spanning over the top of vertical spandrel supports which takes the forces from the deck down onto the arch. Maillart went on to further develop his

FIG 13. The Salginatobel Bridge, Switzerland

ideas by using the stiffness of the bridge deck to restrain the arch laterally and create a very slender deck-stiffened arch bridge. Open-spandrel reinforced-concrete arch bridges are now found over many roads and highways around the world.

The Clyde Arc Finnieston Bridge in Glasgow (Figure 14) is a bowstring arch with a suspended deck. This type of arch is so called because it resembles the shape of a bow with the string between the ends of the bow resisting the outward thrust at its ends just as Brunel did at Windsor. In the next chapter we'll see that it is possible to choose a curve shape so that the force in the arch is almost entirely axial just as for a traditional arch. The difference is that the arch itself is not a series of voussoirs or voussoir substitutes but a continuous beam that can bend should it need to.

FIG 14. The Clyde Arc Finnieston Bridge, Glasgow

The Clyde Arc Bridge was erected in 2006 and was the first over the Clyde since the 1960s and is a major addition to the Glasgow skyline though not trouble-free (see Chapter 7). It is close to the new Glasgow Science Centre and Tower, and the Glasgow Armadillo Exhibition Centre. Building developments on both banks meant that the alignment of the 96-m span bridge had to be skewed across the river—so the locals dubbed it the 'Squinty Bridge'. The designers thought that the usual solution of using two arches would visually conflict. They therefore took the unusual decision to use a single bowstring arch that straddles the bridge deck springing from one side of the deck on one bank to the other side of the deck on the other. The bridge deck is hung from the arch and also works as a structural tie across the bottom of the arch to avoid the need to transfer the horizontal forces to the ground.

Every arch bridge needs firm foundations and they are the hardest to read because you mostly can't actually see them underwater or underground. Nevertheless if we want to be able to read a bridge then we need to appreciate how they work. A foundation must create a smooth transition, allowing the internal forces to flow between the bridge and the ground—it is a link that holds the whole bridge up. If it doesn't work for some reason such that the ground gives way, settles, consolidates, or crumbles, then the bridge may fail.

There are only two types of bridge foundations—although each has many variations. The first is a 'spread' foundation. Here the loads, the forces, from the bridge are spread over an area large enough for the ground to bear them. Spread foundations are rather like the snowshoes worn to prevent you sinking into deep snow—the stereotype is like a tennis racquet—in fact the French term is *raquette à neige*. The principle of a spread foundation applies if you were ever unfortunate enough to find yourself caught in quicksand. The best advice is not to struggle but to lie flat and still on the surface to try to spread your weight. If no one is around to help pull you out, then try to swim—quicksand is liquefied sand and so is much denser than water. Swimming in it will be very hard work but it could save your life.

The second type of bridge foundation uses piles. The principle is similar in that you try to spread the loads to ground that can sustain them. Piles are long slender columns in the ground that transmit the loads to lower depths where they can be carried. For example, if there is bedrock some way down then the piles can be driven or cast so that the loads are taken directly to the rock—they are known as end-bearing piles. However, if there is no bedrock near enough to the surface then the piles may be driven or cast into stiffer ground. The piles are then designed to transmit the load to the ground through the friction between the surfaces of the pile and the ground. A tension pile is one which is used to transmit a pull-out force rather like an anchor. Groups of them are often used to hold down the cables at the end of a suspension bridge. They act a bit like a 'Rawlplug' that you might use to fix a shelf bracket into a brick wall. There the friction that carries the load from the shelf is between the wall and the plug. There is one difference, however, because when you tighten the screw into a 'Rawlplug' you increase the friction—that doesn't happen in a bridge pile.

When bridge engineers talk about soil they don't mean the topsoil in which you grow your garden plants. That soil will not support any weight coming from bridges or any other structures—so it must be removed. To a bridge engineer, soil is the material that you find after the topsoil has been removed. It is largely sand, clay, or a mixture of sediments and deposits that come from the disintegration of rock. As you might imagine the variety of this soil is enormous and for many centuries defied any attempt at scientific understanding. Real soils have solid grains, water, and sometimes air and are usually classified by size as gravel, sand, silt, or clay.

Imagine a heap of dry loose sand—perhaps even one piled up on your favourite beach. If you have ever made such a heap, perhaps as part of a sandcastle, you will know that there is a limit to how steep you can make the sides. No matter how hard you try the sand particles just run down and the slope gets no steeper. Scientists call this slope the 'angle of repose'. You'll see the particles tumbling over each other as they slide because the weight of each one is more than the frictional shear or sliding resistance.

Within the heap the sand particles rub against each other and interlock in a complex way. This creates a shear resistance when any force or load is placed on the sand. For example, if you stand or sit on your heap then the sand grains underneath can sustain that weight as long as the shear resistance of the sand is not exceeded. If you are too heavy then many slips occur along many surfaces within the heap as the interlocking of the grains and the rubbing friction gives way.

Soils vary enormously so the science is very uncertain. There are basically three components: air, water, and solids. The solids are particles. They range in size from clay, where they are the size of dust, through the type of sand you find on a beach, to large boulders. Clay is 'sticky' or cohesive because the particles are so fine that there is an attraction force between them. The particles cannot easily be separated out by filtering or allowing them to settle out—they are colloids where the molecules are larger than those in a solution but smaller then those in a suspension. Milk and paint are common examples of colloids. Sand particles are bigger and so sand is not sticky. The particles slide over each other—but they are rough and so the sliding is resisted by friction between particles.

The shear resistance of dry sand is then almost entirely a friction and interlocking between the grains. Finer grained soils have some 'stickiness' or cohesion so the shear resistance depends on the friction and on cohesion.

Karl von Terzaghi is widely known as the father of soil mechanics and the science of building foundations which we now call geotechnics. He took the ideas of early pioneers, like Frenchman Charles-Augustin de Coulomb who published a theory of earth pressure in 1773. When Terzaghi was thinking about this problem in the 1930s, he realized that one key idea was missing—the pressure between the soil particles in contact with each other. He called this the effective stress. Terzaghi's breakthrough was to realize that this determines the ability of soil to resist load. But he also realized that this pressure depends on the pressure in the water in the soil. He called this the pore water pressure because it exists in the pores of the soil. He defined the total stress as the stress applied to the soil and defined the effective stress as the difference between the total stress and the pore water pressure. In quicksand, for example, all contact between the soil particles has been lost—the effective stress is zero, the water dominates and the quicksand flows just like water—but a very dense water. Terzaghi laid the scientific and engineering foundations for the building of bridge foundations.

So how can we summarize what we have learned about arch bridges? Stone arches are naturally beautiful—the curve of the voussoirs has inspired many artists. Stone bridges are made of natural materials and so are 'of the earth' from which they spring. Stone that has weathered through geological time is durable and long-lasting. This makes arch bridges reassuring symbols of solidity and dependability.

Once a stone arch bridge is up it will stay in place forever as long as foundations don't move. As long as the thrust line of the arch stays within the lines of the voussoirs then the bridge is safe. It is only when the thrust line moves outside of the line of the voussoirs that the arch begins to crack and deform, and when four or more hinges form then the entire structure becomes a mechanism and collapses.

From the cast iron of Ironbridge to the steel of the Clyde Arc Finnieston Bridge, new materials and new ideas have enabled bridge builders to build innovatively. Arches are now built of trusses or beams with cable suspension structures too. To read them in more detail we now need to look at other structural forms. In the next chapter, we will consider the bending of beams.

3

BENDING IT
Bridges Need Strong Structure

The old London Bridge and the Ponte Vecchio in Florence are perfect historical examples of bridges as natural meeting places. Over the centuries, travellers, dreamers, idlers, traders, pilgrims, lovers, diplomats, and soldiers have used bridges as focal points to meet, talk, gossip, trade, worship, find romance, negotiate, and even fight. Many battles have been fought on and around bridges. In modern warfare, strike aircraft target bridges to cause maximum disruption to everyday life. Bridges are an essential part of the structure of our collective human identity. People need people and bridges link people. Bridges are essential gateways across natural frontiers such as rivers. Where there are no bridges, marked differences evolve between modes of living, culture, and customs on each side. So the story of bridges is the story of the fulfilling of a basic human need—to cross barriers to link people.

The first regular river crossing points were the fords that became part of well-used tracks and trade routes. Some fords were marked with a prepared approach with a firmed-up or hardened riverbed. Many English towns such as Oxford owe their names to the original ford. Indeed fords were more common than bridges in early place names. A small town was called Granta bryg when a bridge replaced the ford on the River Granta. Later Granta became Cam and the town became Cambridge. Brigstowe, meaning 'place by the bridge', in 1051 became Bristol. In the south of England the name 'Stoke' is frequently associated with a crossing place as it referred to the staking which firmed up the track over boggy ground by supporting wattles and faggots (bundles of flexible sticks, branches, and twigs). When the ford was flooded or impassable, because the water was too deep or the current too strong, then ferries began to find good business.

Stepping stones often occurred naturally. Often new stones were added or indeed whole new crossings created. A simple tree trunk or log might span a stream but when the gap was too wide timbers or stone slabs were laid between stepping stones. These bridges became known as 'clapper bridges' and the stepping stones became the first bridge piers or foundations. The word clapper, in this context, means plank but more recently has tended to be used only for dry stone bridges with flat stone slabs. Many still exist, as at Tarr Steps on Exmoor, UK (see Figure 15). This clapper bridge is about 155 m long with 17 spans of stone slabs about 1.5 m wide laid on stone piers up to about 4.5 m wide. The biggest surviving clapper bridge, the Anping Bridge, was built in China more than 800 years ago. It crosses a 2-km stretch of sea and feels timeless and tranquil according to the reports of some visitors.[1]

I mentioned that bridges are important in warfare. When armies needed to cross deep and wide stretches of water they used boats, tied together, to form a floating bridge. Herodotus reported that Xerxes, King of Persia, took his armies across the Hellespont in 480 BC using a floating bridge. Herodotus wrote:

FIG 15. Clapper Bridge, Tarr Steps, Exmoor, UK

Penteconters and triremes[2] were lashed together to support the bridges—360 for the one on the Black Sea side, and 314 for the other. They were moored slantwise to the Black Sea and at right angles to the Hellespont, in order to lessen the strain on the cables. Specially heavy anchors were laid out both upstream and downstream—those to the eastward to hold the vessels against winds blowing down the straits from the direction of the Black Sea, those on the other side, to the westward and towards the Aegean, to take the strain when it blew from the west and south. Gaps were left in three places to allow any boats that might wish to do so to pass in or out of the Black Sea.

Once the vessels were in position, the cables were hauled taut by wooden winches ashore. This time the two sorts of cable were not used separately for each bridge, but both bridges had two flax cables and four papyrus ones. The flax and papyrus cables were of the same thickness and quality, but the flax was the heavier—half a fathom of it weighed 114 lb. The next operation was to cut planks equal in length to the width of the floats, lay them edge to edge over the taught cables, and then bind them together on their upper surface. That done, brushwood was put on top and spread evenly, with a layer of soil, trodden hard, over all. Finally a paling [a fence] was constructed along each side, high enough to prevent horses and mules from seeing over and taking fright at the water.[3]

You can see that Xerxes went to some considerable trouble to build his bridges and so he wasn't very pleased when things went wrong. When a storm smashed up one of his bridges he ordered that the water be severely reprimanded by giving it three hundred lashes and have pairs of fetters thrown into it. He then gave orders that 'the men responsible for building the bridges should have their heads cut off.'

It is fortunate for modern pontoon bridge builders that they aren't dealt with so severely, because a number of pontoon bridges have suffered from problems. In 1980 my friend Colin Brown drove me over the Lacey V. Murrow Memorial Bridge near Seattle, the first large pontoon bridge I had ever seen. Colin had settled into an American way of life in the 1960s without ever losing his essential Englishness. He told me that the original bridge was opened in 1940 and directly connected Seattle to Mercer Island. It was called the Lake Washington Floating Bridge but was renamed in 1967 in honour of the highways director of Washington State Lacey Murrow (brother of Ed Murrow, the well-known American broadcaster). The original bridge consisted of 25 concrete box pontoons—essentially the modern equivalent of what Herodotus described—but with specially made concrete boats. Some of the pontoon boats were as long as 115 m and 18 m wide and all were anchored to the seabed. At the eastern end specially framed pontoons could be retracted longitudinally to create what is known as a draw span for ships to pass through. Unfortunately, the bridge sank in November 1990 during a US$35 million renovation project to widen and resurface the highway. High-pressure water was being used to remove unwanted material but it was too contaminated for the lake. The project engineers therefore decided it could be stored temporarily in the pontoons. Consequently the watertight hatchways were left open during a Thanksgiving Day holiday. During the holiday weekend there was a large storm and some of the pontoons were filled with rain and lake water. As soon as the problem was realized the workers started to pump out the water. But it was too late—one pontoon sank and dragged down the rest as they were cabled together. A new bridge was built in 1993.

Floating bridges work best when a long crossing is required (a few miles) over deep water (over 200 ft), so they are very suitable for the waters around Seattle. The Evergreen Point Floating Bridge was the second to be built in the area and is the longest in the world at 2,310 m, taking the Washington State Route 520 across Lake Washington from Seattle to Medina. The Hood Canal Bridge was the third floating bridge about 40 miles NW of Seattle. It was opened in 1961 but failed in 1979 in a similar way to the Murrow Bridge. The floating bridge across Lake Okanagan in British Columbia, Canada is a critical lifeline structure linking Kelowna and Westbank. Norway has two floating bridges, in BergsCysund (1992) and Nordhordland (1994). Unfortunately, floating bridges have a history of failing too frequently but they are still viable under the right conditions. At least now we don't cut off the heads of the designers.

Floating bridges are rare but, like the more conventional bridges we are going to look at for the rest of this chapter, they are simply long beams that flex. To read any beam bridge we need to understand the flow of the internal forces as it curves from its initially straight form, just as we did for the thrust line within an arch in the previous chapter.

There are just three chapters in the story of a beam bridge. The first contains the main beams that span along the length of the bridge from one side to the other, together with any crossbeams that hold the main beams together. It also includes the pontoons of a floating bridge. The second chapter contains the bridge deck on which people walk or vehicles travel as well as handrails and side barriers. The third chapter is about the foundations that, as always, are crucial. Floating bridges are supported by the water but also must be anchored. The grammar of beam bridges—the science of the flow of forces—will be easier to understand through a little basic algebra but you can skip over the mathematical details and still 'get a feel' for how beams work.

In essence, beam bridges are really just very big manufactured complex planks. They are structural workhorses—they just get on with the job of responding to whatever is asked of them—bending this way and that way to cope with the loads. But we must be careful because in certain circumstances beams can fail in surprising ways as we will discover later. Of course only the simplest plank can work singly. For any sizable bridge there must be a network of beams working together—some along the length of the bridge and some spanning across the width.

The ways they work together depends totally on the context, on the particular situation for which they were designed, were built, and are being used or operated.

The chapters for a simple footbridge, perhaps over a small stream, are relatively easy to see but they may not be totally distinct. For example, the cross plank beams may also be the decking. The handrails of a beam bridge are not normally part of the structure of the bridge but of course they must have sufficient strength and robustness to resist anyone leaning on them.

A major highway bridge is more complex but you can still spot the chapters. The best way is to stand beneath the bridge and look up at its understructure There are many variations but most are on the same theme. Firstly, as I have said, there will be the main beams spanning between the supports. Sometimes there is just one single large beam—usually in the form of a box section. More commonly there will be a series of two or more main beams sitting side by side. For example, Figure 16 shows an underside view of the two massive concrete box beams for the English-side approach spans of the Second Severn Crossing that takes the M4 over the Severn estuary near Bristol. From the side view it is clear that the beams are shaped. There are two reasons for this: the first is aesthetic and the second structural. Not all big highway bridges have a curved elevation but many do and they often look much better than constant-depth girders. Beam bridges are often deeper at the supports because the internal forces are bigger as we will see in a moment. In Figure 16 you can see the smaller secondary cross beams that span between the main beams (and the smaller gantries for travelling maintenance cradles). Some bridges have tertiary beams sitting on the secondary crossbeams.

FIG 16. The underside of the Second Severn Crossing approach spans

The bridge deck of a highway bridge is usually a concrete slab or a steel reinforced plate and may also be the top flange of the main beam. On top of the bridge deck will be a wearing surface usually of asphalt on which the vehicles travel. The deck will support all the other normal street furniture such as drains, streetlights, crash and noise barriers, and handrails.

The main beams may span simply between two supports or continuously over several as in Figure 16. The junction between the bridge girders and the foundations is usually designed to be quite distinct so that the forces from the bridge are taken to the ground in a fairly precise way—you can see the bearings in the photograph.

Why are beams shaped the way they are and is there a difference between a beam and a girder? I referred to some of the many varieties of manufactured beams briefly in Chapter 1 and Figure 3. There are two basic forms: they have either a closed or an open cross section. A closed cross section has a contained inner space like a rectangle or a triangle—a common example is a box beam or a tube. An open section has no contained inner space—a common example is an I-beam. A very large beam is called a girder though the terms are often used interchangeably.

An open section beam can have a round cross section like a tree trunk or be rectangular like a plank. However, for larger beams and girders a cross section in the form of a capital letter I or the letter H turned on its side is efficient. This shape has a vertical part, the centre of the I and the rotated H, which is called the web. The two horizontal parts at the top and bottom of the I and the rotated H are called the flanges. Another way of thinking about this is to see the beam as a rectangle with two smaller equal rectangular 'bites' taken out of each side. These bites can be taken because the material removed is not doing much. The internal stresses that resist the bending are concentrated in the flanges of the I-beam and the stresses that resist the shear are almost all in the web. So an I-beam is very much lighter than a rectangular beam for much the same strength.

A box girder typically has a rectangular or trapezoidal closed cross section. A trapezium has two parallel sides, which are the top and the bottom flanges, with sides that slope outwards to make an efficient aerodynamic shape. Such a beam is almost like an aeroplane wing but with the difference that the shape is designed to *reduce* the uplift effect of the wind not to maximize it as when an aeroplane takes off. Box girders are normally made of concrete—as in Figure 16 the approach spans of the Second Severn Crossing and steel for the central cable stayed section of that bridge—but there are examples of box girders built of timber or even plastic. The closed box shape has a good resistance to torsion as required, for example, when a bridge is curved in plan.

In the previous chapter we saw that arch bridges work by transferring a flow of internal compression along a thrust line. Beams work mainly by transferring a flow of internal forces of bending moment and shear along and across the beam. You can get a feel for these forces when you bend forwards or bend down. You sense your back being pulled in tension and your tummy being squashed in compression.

Beams can also transfer internal compression and so are used as arches as already mentioned at Salginatobel Bridge (Figure 13) and the Clyde Arc Finnieston Bridge (Figure 14). So beams have three degrees of freedom (see Chapter 1)—movement in two directions and a rotation. Any restrained up-down movement creates shear, restrained longitudinal movement causes tension or compression, and restrained rotation causes bending moment. We will look at beams as arches in more detail later in this chapter but for now let's just focus on the main way in which they work, i.e. bending and shear. It isn't obvious that the stone slabs of the clapper bridges and the massive beams of modern highway bridges do bend. The movements are so slight that you can't detect them without special measuring instruments. However, you can often feel a vibration on a big bridge as a heavy truck passes. To see bending with the naked eye we need something much more flexible. A strip of wood or plastic, like a ruler, will do the job perfectly.

Imagine resting the ruler on two supports (two thick books, say), one at each end. Now press down on the middle with a

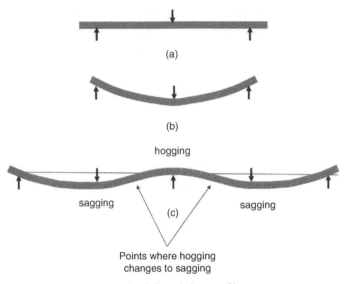

FIG 17. The deflected shapes of beams

finger. The ruler will bend quite easily and you can see the curve of the bent shape—see Figures 17a and 17b. In the diagram the downward arrows are the vector forces of your finger pressing down and the upward arrows are the consequent reactions from the books resisting your finger. The harder you push the more the beam sags and the more the curvature of the bend. The bent ruler is an example of the simplest of all beam structures— engineers call it a simply supported beam for obvious reasons. The clapper bridge stone slab is also simply supported. As we shall see in more detail in a moment the degree of curvature creates strains as the bottom of the beam stretches and the top of the beam compresses. As always the consequent forces and stresses must be resisted in the three ways of being strong: tension, compression, and shear.

But before we get to that, let's insert an extra support at the centre of our simply supported beam. In effect we will now have a beam with two spans and three supports. Engineers call that a continuous beam—a two-span continuous beam. We could imagine creating a three-, four-, or even five-span continuous beam by having the appropriate number of internal supports.

To demonstrate this we'll need a longer ruler or strip of wood, perhaps about a metre long, supported at its ends and in the middle as in Figure 17c. We then press down with one finger in the middle of each span. As long as our beam doesn't crack or break, it will bend into a continuous curve. The shape is a little more complicated than the simply supported beam but we can work it out quite easily. Start at either end and look at the shape—the beam 'sags' down towards your first finger. Then it has to bend back to go over the internal support. Engineers call this bending back 'hogging'—as distinct from 'sagging'. As we follow the shape into the second span the beam continues to hog but as we approach the second finger it starts to sag again and then curves back to the outer support. Since the two spans are the same their shapes are symmetrical.

So far we have made two very important assumptions that we need to be clear about. First, we have presumed that the beams can rotate freely at the end supports. Whilst this is often the case in a real bridge—it is not always so. Fixed or 'encastré' supports hold the beam ends in a tight clamp so that they can't rotate—in other words a degree of freedom is restrained and so an internal force is created—a bending moment. An example is a cantilever such as a diving board which is encastré at one end and entirely free at the other. If one end of a simply supported beam was encastré, i.e. held rigidly horizontal, then the beam's deflected shape would have to bend back on itself. In other words there would be a hogging bending moment.

Our second assumption is that the deflections are very small— I have exaggerated them in the diagram so you can see what is happening—if they became that large in reality then the beam would behave differently. We will soon be making an important third assumption, which is that the beam is made of material which behaves as though it is linear elastic as defined in Chapter 1. We'll examine the implications of these last two assumptions in more detail in Chapter 6.

So what is going on inside the beam as it sags and hogs? In Chapter 1 we distinguished between internal and external forces for a tug of war rope. We said that the internal tension force was axial, i.e. along the length of the rope so there is only one degree of freedom. When we made our imaginary cut through the rope the internal force was the same on every individual bit, or element, of the cross section. This meant that the stresses (the forces on any small bit of the rope) were the same across the cut and we only needed one person to hold each cut end to simulate the internal tension force. The situation within beams is similar but just a bit more complicated—we'll find that we need three people—one for each degree of freedom in an 'internal force team'.

Let's go back to the simply supported single span beam in Figure 17a. We have said that if we push down in the middle it will bend as shown in Figure 17b. Now let's make an imaginary cut through the centre of the beam in exactly the same way as we did for the tug-of-war rope to expose a beam cross section.

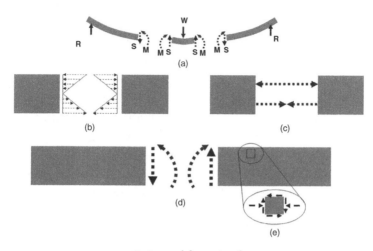

FIG 18. Internal forces in a beam

Indeed let's go one stage further as in Figure 18a where I have made two cuts to expose a small piece or element of the beam. This means we will require two internal force teams—one at each cut. Now at each cut there are two internal forces required to maintain equilibrium. I have labelled the first one as S. This is a vertical internal shear force needed to keep the vertical forces on any piece of the beam in balance. So the job of the first member of our internal force team is to take hold of both ends of each cut and lift up one side and push down on the other. The second internal force is labelled M and is a force called a bending moment that is needed to balance the tendency of any piece of the beam to turn or rotate. We'll need the other two members of our internal force team to resist the bending moment. Both of these internal forces, shear force and bending moment, need a bit more explanation.

When we cut the tug of war rope in Chapter 1 the internal forces and stresses didn't vary over the cross section. A beam is more complicated because the internal forces and stresses do vary over the cross section.

As the beam sags, each small element on the bottom of the beam is being pulled or stretched—in tension. At the same time each element at the top of the beam is being squashed or compressed.[4] Therefore the stresses in the cross section change from tension at the bottom to compression at the top. These changes are through the depth of the cross section—there is no variation across the width.[5] As they change through the depth there is a point where the stress changes from tension to compression. At that point the stress is zero. We call it the 'neutral axis' of the cross section.

We have said the stresses vary up and down the cross section. But how do they vary? For many materials we can assume that the stress changes as a straight line as shown in Figure 18b.

You'll recall the stresses are vectors and therefore can be represented by arrows. In Figure 18b the compression stresses at each level through the top of the depth of the beam are depicted by the arrows pointing into the beam face. Similarly the tensile stresses in the lower half of the beam are the arrows pointing away.

In Figures 18b and 18c, the vertical shear forces have been omitted to simplify the diagram. For any cross section of beam symmetrical about its middle, the neutral axis is at the centre as shown. For more complex cross sections, for beams made of more than one material or for loads that cause very high stresses and strains, the neutral axis may not be in the middle and the variation of stresses may not be linear.

If we were to take cuts through cross sections along the rest of the length of the beam then we would find different patterns of stress changes. For example, over the internal support in Figure 17c we have the exact opposite situation to that at the centre of the span because here the top of the beam is in tension and the bottom is in compression as the beam hogs.

Because of these changes along the length of the beam the amount or degree of bending—the curvature—changes along the length of the beam. At some point along the length the sagging changes to hogging and the hogging to sagging. We call these 'points of contraflexure' because at this point the beam is not bending at all; it is just at the point of changing from one direction to the other. We'll make a diagram of these changes in a moment (Figures 19, 20).

Let's now focus on any cross section through the beam as shown in Figure 18b. The total force due to the stresses above the neutral axis must balance those below. This must be so to maintain horizontal equilibrium. However, because one is tension and the other is compression they create a turning effect.

In Figure 18c I have replaced the distributed stresses above the neutral axis with a single arrow. This arrow represents the sum of all of the distributed stresses as a total force in compression in the top half of the cross section. Likewise the lower arrow represents the total force in tension in the bottom half. The turning effect is now clear as the top force pushes into the cross section and the bottom force pulls it out. So now we can ask the two remaining members of our internal force team to take their places. One must pull—just as for the tug of war—against the bottom tension force in Figure 18c. The other team member must push against the upper compression force.

With these internal forces in place every little piece of material in the beam would rotate—if it were free to do so—but it can't because the rest of the beam stops it—or rather restrains it. Each element rotates but not as much as it would if it were entirely free. The internal force that creates this restraint is called the bending moment and is now being resisted by the two latest members of our internal force team. Conventionally the two forces resisting the turning are shown using the curved arrows M in Figures 18a and 18d. To present the complete picture the shear force has also been reintroduced in Figure 18d.

Unfortunately this isn't the end of the story. We need now to look at the forces and stresses on a small element at the top part of the beam in Figure 18d. This element is being compressed by the bending. When we isolate the element as in Figure 18e we can see that not only is it being compressed it is also being sheared.

If we were to do the same thing in the lower part of the beam we would find that an isolated element there would be stretched in tension as well as being sheared.

Even more puzzling perhaps is that a horizontal shear force as well as a vertical one is shown. The reason is that, if it were free to do so, the bending of the beam would make our element slip along the surface between it and the rest of the beam. Because it can't slip we get an internal shear force.

Let's imagine our element extended to the full length of the beam. We will also make it a bit deeper, say a third of the total depth, and disconnect it from the rest of the beam. Now we have made three flexible layers of simply supported beams—in effect three flexible rulers—sitting on top of each other. Imagine pressing down on the middle of these three stacked beams but remembering as we do that they are free to slide over each other. Each one will bend to a slightly different curvature—just like a leaf spring. You will be able to see the relative movement at the ends. Now if we stick the three rulers back together again so they can't slip (like the blocks we looked at in Chapter 1) then there will be no relative movement at the ends because the slipping is constrained. An internal horizontal shear force between the layers will be created because the layers want to slip but can't.

There are two types of bending moment and shear force. The first type, called the applied bending moment and applied shear force, is created as the external loads try to turn or rotate and shear each little piece of the beam. This is what we have been discussing so far. The second type, called the resisting bending moment and resisting shear, represents the strength of the beam. The values depend on the shape of the beam and the material from which it is made. The resisting moment and shear force must be bigger than the applied moment and shear force if the beam is to be safe—but we will return to how we achieve this in Chapter 6.

In order to keep the beam safe, bridge builders need to understand how the applied bending moments and shear force vary as they flow through a beam. Therefore they draw diagrams such as those in Figures 19 and 20 to help identify where the internal forces are the largest.

Figure 19a is the simply supported beam again, but this time with all of the dimensions and external forces identified. The diagram shows how the bending moment and shear forces change along the beam as a straight line in Figures 19d and 19e. You can see that the largest bending moment occurs in the middle.

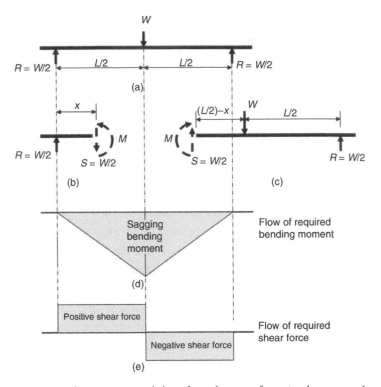

FIG 19. Bending moment and shear force diagrams for a simply supported beam

Let's understand why, by looking at how the external forces balance in Figure 19a and then at how the internal forces balance in Figures 19b and 19c.

The applied vertical load is W and is acting right in the middle of the span of distance L—so the situation is symmetrical. It follows therefore that the reactions R must be equal. As the reactions must balance W they must both be $W/2$.

Now let's focus on the left-hand piece of the beam in Figure 19b. Instead of a cut exposing the internal forces at the centre we make the cut at some distance x from the left-hand support. In order for this piece of beam to be in equilibrium then $S = R = W/2$. The shear force S is half the load W.

However, S and R together would turn this piece of beam if it were free to turn but the rest of the beam stops it. Consequently, an internal balancing moment of M is created.

Recall that a moment is the turning effect of a force about a point. It is the force times the shortest distance to the point.

In our case the force is R and the distance is between R and S and is x.

Consequently $M =$ the force R times the distance x.

Since $R = W/2$ then $M = (W/2) \times x = Wx/2$.

The largest M occurs when $x = L/2$ and

$$M = (W/2) \times (L/2) = WL/4$$

If you do the same calculation from the right-hand end on Figure 19c you will get a mirror image of the same answer. The shear force remains constant but changes direction at centre span.

Whether S or M is shown positive or negative in Figure 19 is purely a matter of convention. Here sagging is drawn negative.

The maximum hogging and sagging bending moments for the two-span beam in Figure 20 occur under the loads and over the central support.

Beams bend but they can also twist. Just as a bending moment is associated with rotation so a torsion or torque is associated with twisting. An example of a twisting beam is a motor drive

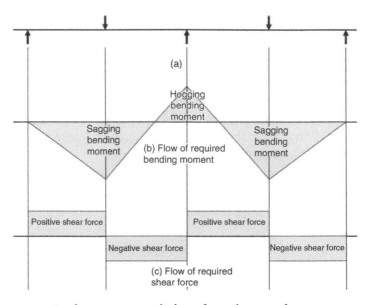

FIG 20. Bending moment and shear force diagrams for a two-span continuous beam

shaft in an engine. In our discussion so far we have assumed that the applied forces bear straight down on the centreline of the cross section of the beam. We haven't allowed for a loading that is off-centre of the cross section. An eccentric load of that kind would twist the beam. For example, if all of the pedestrians on a footbridge decide to stand on one side of the bridge (for example, to watch a fireworks display) and there is no one on the other side then the bridge will be twisted. We are not going to consider this problem in any detail except to say that twisting results in a different combination of tensile, compressive, and shear stresses but with all of the six degrees of freedom we introduced in Chapter 1—movement in three dimensions and rotation in each of the three planes. However, there are still only three ways in which the beam must be strong—tension, compression, and shear.

As well as torsion and twisting there are many other important factors that bridge builders must also consider—two of the most important are buckling (see later this chapter) and fatigue (see Chapter 5).

So far we have restricted ourselves to beams that are wide relative to their depth—like our ruler. However, the larger the distance between the net tensile and compressive forces in Figure 18d, the larger the turning effect. So a deeper beam has more capacity to resist bending than a shallower one. In practice therefore most (but not all) real beams tend to be deep and narrow rather than wide and flat like our ruler. Unfortunately there is a strong downside. As we saw in Chapter 1 when a slender element is compressed it tends to move sideways—it buckles. The compressive stresses on one side of the neutral axis of a slender beam can cause that part of the beam to buckle. Since this does not happen on the tension side the beam tends to twist. The tension side remains straight but the compression side moves sideways—an effect known as lateral torsional buckling. Bridge builders can prevent this by restraining the compression flange. One simple technique, when two beams are sitting side by side, is to brace them together so they support each other laterally. Lateral torsional buckling can severely reduce the carrying capacity of a beam as the story of the Dee Bridge illustrates.

The Dee Bridge was part of the Chester–Holyhead railway. On 24 May 1847 it collapsed as a train was crossing and five people were killed. Just as new materials changed arch bridges so cast iron, wrought iron, steel, and concrete changed beam bridges.

The Dee Bridge with its three 29.8-m spans was one of the first cast iron bridges to catch the spotlight. It was designed by Robert Stephenson and completed in 1846. The two parallel cast-iron girders were both made of three very large castings dovetailed together and strengthened by wrought iron bars.

About 60 similar structures had been built successfully between 1831 and 1847 but the Dee Bridge was the largest. They were all designed using a formula developed by Eaton Hodgkinson based on a number of tests on almost perfectly straight beams of up to 3 m span—about a tenth of the spans of the Dee bridge. Casting technology of the time was at its limits so for the Dee Bridge Robert Stephenson had to accept an out-of-straightness of up to 76 mm. It was a decision he didn't realize would turn out to be of immense importance.[6]

The resulting commission of inquiry in 1849 into the use of iron bridges was very thorough. They examined everything they could think of—'concussions, vibrations, torsions and momentary pressures of enormous magnitude, produced by the rapid and repeated passage of heavy trains'.[7] They did experiments on impact and fatigue, including two full-scale tests on the Ewell Bridge on the Croydon and Epsom Line and the Godstone Bridge on the South Eastern Line. The experts of the day such as Brunel, Stephenson, Fairbairn, Locke, Cubitt, Hawkshaw, Fox, and Barlow were questioned closely. But nowhere in the whole investigation is there any mention of the modern explanations of why the bridge collapsed—lateral torsional buckling of the compression flange or metal fatigue. Why? Because at that time no one knew what lateral torsional buckling was and only a few had seen metal fatigue in railway axles. These were phenomena that had not been identified.[8] The question 'How can we know what we don't know?' is a fundamental one for any modern risk analysis—we'll return to it in Chapter 6.

Robert Stephenson and I. K. Brunel were friends and often consulted each other even though they were also rivals. They would visit each other's projects to give support. Robert was born on 16 October 1803. He went to the local village school in Killingworth near Newcastle upon Tyne. The growing success of his father, George Stephenson, enabled a private education at the Bruce Academy in Newcastle. By 1819 Robert was an apprentice at Killingworth Colliery. After three years he helped his father survey the Stockton and Darlington railway line. In 1823 Robert, George, and Edward Pease formed the world's first locomotive building company. Robert went to Colombia in South America in 1824 and worked at gold and silver mines. Three years later he returned to England and began work on the Rocket locomotive. In 1833 he was appointed chief engineer of the London and Birmingham line. This was the first railway into London and involved solving some difficult engineering problems including cuttings and tunnels. He became Conservative MP for Whitby in 1847. He was President of the Institution of Civil Engineers for two years from 1855. His remains are buried in Westminster Abbey.

Whilst Brunel was using wrought iron for the Saltash Bridge (Chapter 2) Stephenson was using it for one of the first major box girder bridges, the Britannia Bridge that crosses the Menai Straits between the island of Anglesey and the Welsh mainland. Unfortunately, the bridge was destroyed in May 1970 when five teenagers playing with flames accidentally started a major fire. It quickly spread to the timbers and the wrought iron of the bridge which were both coated with tar. The fire brigade fought for nine hours but the bridge was burnt out. Robert Stephenson's 120-year-old bridge was almost totally destroyed in a single night.

Like the Dee Bridge, the Britannia Bridge was part of the rapid development of the railways in the early to middle part of the nineteenth century. Something like 30,000 bridges were built between 1830 and 1868. The Chester and Holyhead Railway was authorized by Parliament in 1844 with a bridge over the Menai Straits. In 1826 Thomas Telford had completed a suspension road bridge across the straits but this type of bridge is normally too flexible for railways. The roadway had to be 103 feet above the water across the entire span to allow for navigation. This meant that an arch bridge wasn't practical—a new idea was needed.

So Robert Stephenson decided to build a wrought iron tubular box beam bridge with the trains running straight through the middle. The two main spans were to be 140 m sitting on masonry piers with side spans of 70 m. The centre pier was to be built on Britannia Rock. At that time the longest wrought iron span was 9.6 m, so Robert knew he had a challenge. There were three novel aspects. Firstly previous bridges of this length of span had been suspension or arch type. Although cast iron beams were being used extensively there was little experience of wrought iron and that mostly in shipbuilding. Finally previous beams were I- or T-shaped—they were not tubular.

He therefore asked William Fairbairn to help him. William was a shipbuilder and iron fabricator and also an old friend of his father. Fairbairn realized that some detailed work was needed and he enlisted further help from Eaton Hodgkinson, a man who knew some of the latest theories in the strength of materials. Hodgkinson didn't believe that Stephenson's tubes could be made stiff enough and that suspension chains would be needed similar to those later used by Brunel on his Chepstow Bridge in 1852 (see Chapter 2).

Fairbairn didn't agree so he set about proving it by experiments at his shipbuilding plant at Millwall in London. His preliminary tests were on tubular beams with circular, elliptical, and rectangular cross section. He varied the size of the section, the thickness of the plate, and the length of the span—up to 10 m. The tubes were riveted wrought iron following the practice of boilermakers. They were simply supported and loaded at midspan. The load was increased steadily until the beam failed. Fairburn quickly found that boiler-making methods weren't good enough for bridge beams. More importantly he realized that the beams nearly always failed by buckling and wrinkling of the plates in the top side of the tube. Fairbairn wrote that the results were completely unanticipated—they were 'anomalous to our preconceived notions of the strength of materials, and totally different to anything yet exhibited in any previous research'.[9]

Fairburn, Stephenson, and Hodgkinson decided to opt for the rectangular section because they felt that this shape was one they could understand and deal with best. But they knew they had to sort out the buckling of the top plate of the box beam. They realized that it wasn't the strength of the material that was important—rather it was its shape. They knew they had to find a form for the top plate that didn't buckle so easily. They decided to try replacing the plate with a series of smaller tubes. So in July 1846 they tested a (1/6) scale model of the final bridge beams with six square cells in the top of the beam. Since the beam was an approximately 23-m span, the tests were expensive and the men needed to get as much information as possible. So they did a test, repaired the bit that failed, then conducted another—a classic case of trial and error. By the sixth and final test in April 1847 they had developed a beam that failed in the top by buckling at almost the same time as it failed in tension at the bottom. They had increased the failure load by a factor of 2.5 using only 20% extra material.

Other modifications were required. For example, the bottom plate had to be strengthened to take the wheel loads for the trains passing through the box. The sides of the tubes were strengthened to take the shear forces. Stephenson decided to make the beam continuous over the support so it became the largest beam ever at 457 m long. A good measure of the work involved is the number of rivets—2,190,100 of them! Of course the beam couldn't be made in one piece so smaller sections were floated out on pontoons and hoisted to their final positions using hydraulic presses. The boxes were lifted 2 m at a time and then timber and masonry underpinning was built up beneath the tubes. It was fortunate that Stephenson had planned very carefully because when the first tube had been lifted about 24 ft the cast-iron cylinder of the large press burst violently and the beam fell. There was some damage but the timber packing saved the day and no one was hurt.

Once in place on top of the piers and already deflecting under their own weight, the giant tubes were joined together. Stephenson knew that the stresses due to bending could be reduced by controlling the joining-up process. So he measured the deflections during the erection and jacked up the sections to try to equalize the strains at mid-span with those over the supports. In doing this he was surprised to find a significant movement of 60 mm horizontally and vertically from day to night simply due to temperature changes.

The Britannia Bridge became one of the most successful railway bridges in the UK. Stephenson went on to design the High Level Bridge in Newcastle upon Tyne. His work influenced Brunel's Royal Albert Bridge across the River Tamar at Saltash (Chapter 2). After the fire the bridge was rebuilt with the deck on two levels (rail and road) supported by arches and opened in 1972.

Eventually cast and wrought iron were replaced by concrete and steel. Robert Maillart (Chapter 2) took advantage of the new ideas of reinforced concrete to make arches in which the arch ribs are really curved beams. Concrete is weak in tension so if we make a simply supported beam in reinforced concrete then the steel reinforcing bars must be placed in the bottom of the beam where the bending induces tension.

As we said earlier a continuous beam over two or more spans hogs over its central supports. It therefore has tensile stresses in the top—so that is where the steel reinforcement must be. At the point of contraflexure where sagging changes to hogging the steel at the bottom will overlap the steel at the top just to make sure that the internal forces are transmitted properly.

Shear stresses in a beam tend to make an element deform into a lozenge or diamond shape. You'll recall from Chapter 1 that this means that one diagonal will get longer and one shorter. As this tendency to lozenge is being resisted by the rest of the beam a tensile force will develop in the diagonal that is stretching. Since that is bad news for the weak concrete, bridge builders reinforce it with steel bars. Normally they bend bars upwards at an angle of about 45° at the ends of the beam where the shear force is the greatest. This simply helps to strengthen the beam along the diagonal. Concrete beams reinforced properly in this way can resist significant bending moments.

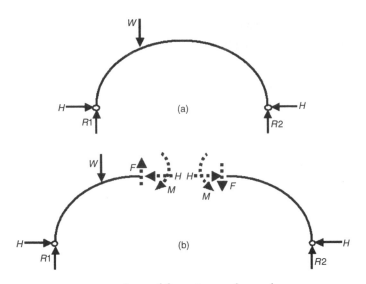

FIG 21. Internal forces in a modern arch

The designers of an arch bridge with a continuous arch rib, whether made of reinforced concrete or steel, can choose a shape that reduces the bending moments significantly. They usually choose a shape where the bending moments due to applied loads are balanced by the bending moments present because of the shape of the arch. You can see this in Figure 21a where an arch rib is carrying a single load W. Note that the ends of the arch are hinged so that they are free to rotate. This type of arch is called a two-pinned arch. Going back to the beam ruler we looked at earlier you might imagine an arch rib as a curved ruler instead of a flat straight one.

If you push down on this curved ruler then the bottom ends will want to spread out. The foundations resist by pushing back. W is shown as the pushing force as in the diagram and H is the foundation pushing back. Unfortunately, H is not easy to calculate.

The situation this time is not symmetrical. The vertical reactions—R_1 and R_2—will not be equal. In Figure 21b the arch has been cut through to expose the internal forces at the top. The main difference from the simple flat beam we looked at earlier is that there is now an internal thrust due to the force H.

If a cut were to be made anywhere else along the arch the internal forces would be exposed in just the same way. They would have different values of bending moment and shear force but the horizontal force would be the same since there are no other horizontal forces on the arch.

Imagine doing this at a quarter point, for example. Then imagine resolving the horizontal force H and the vertical shear at the quarter point into two forces. One would be along the arch and the other at right angles to it. The force along the arch is the axial thrust at the cut and the other is the shear normal to the line of the arch. The bending moment is unchanged. Of course as we go around the arch the angle of this axial thrust will change because it is always along the line of the arch.

This changing thrust around the line of the arch is equivalent to the thrust line of the masonry arch in Chapter 2. The main difference is that our arch now has internal shear forces and bending moments as well.

The external forces R_1, R_2, and W are exactly as for a simply supported beam. However, we now have H providing a turning effect, or moment, in the opposite sense. The effect of H therefore is to reduce the bending moment M in the arch from that of a beam.

You can see this when you spot that R_1 acts in a clockwise direction about the cut cross section and W acts anticlockwise as for the flat beam. However, H acts anticlockwise too and this reduces the total bending moment M. Indeed if we organize the shape of the arch carefully we can make M become zero.

In Figure 21b if R_2 multiplied by half the span equals H times the height of the arch then M will be zero.

Bridge builders often choose a parabolic shape for a beam arch rib such that the bending moments are zero or very small under the self-weight of the whole bridge. Then the arch rib will only have to resist bending stresses induced by traffic or wind or other hazards that they call live loads. Brunel made his arches rise the same distance as his chains sagged at Saltash. Some arch bridges have a pin at the apex as well as at the foundations. The Sydney Harbour Bridge was under this condition when the two halves first met. If there are more than three pins, then the entire arch becomes a mechanism and collapses.

Just as Maillart was exploiting reinforced concrete for this type of arch rib another Frenchman, Eugène Freyssinet, was developing a newer and more advanced idea—prestressed concrete. Freyssinet was born at Objat, Corrèze, France. He designed several bridges before the First World War and served in the French Army. Hopkins wrote that Freyssinet's most beautiful bridge is the Pont sur la Marne at Esbly, Seine-et-Marne with a 74-m span, built in 1947.[10] Freyssinet didn't invent the idea of prestressing but he did develop it and use it.

Imagine a horizontal hole right through the length of a concrete beam at the centre of its cross section but near the bottom. Then thread a steel bar with screw threads at its ends through the hole. Tighten up two nuts at each end of the bar until they bear up onto the concrete of the beam. As the nuts are tightened the bar begins to stretch and the bottom of the beam is compressed.

The compression force is off-centre and so the beam begins to bend upwards—to hog. This then 'prestresses' the beam. If a normal downward load is now applied to make the beam sag then the sagging from the load and the hogging from the pre-stressing tend to cancel each other out to some degree. As a consequence the resulting total stresses are smaller. Using this idea the bridge beam designer can create a beam that is smaller and lighter than would otherwise be the case.

Freyssinet experimented with prestressing some early bridges. For example, he left an unconnected gap in the crown of a concrete arch. He inserted jacks and applied a horizontal force to the two halves which lifted them clear of the centring. Before removing the jacks he concreted in the crown leaving a perman-ent set and hence a prestressing. After many experiments of this kind he took out a patent for prestressed concrete in 1928. Freyssinet's key contribution was to recognize that creep of concrete was a problem. Concrete creeps when it is put into compression over time because it relaxes and shortens. This means that the tension in the prestressing bar relaxes and re-duces and is therefore much less effective. He also realized that high-strength prestressing wire was more effective and that he needed to develop anchorages that might be usable in a variety of structures.

Over the years various techniques have been developed for applying the prestress. Forms of prestressing more elaborate than the single bar described above have been developed. One common technique is to use many thin wires in ducts along the length of the concrete beam but which are curved in side eleva-tion, i.e. when looked at from the side. As well as reducing longitudinal stresses the wires also assist in resisting shear.

Sometimes the wires act quite separately from the concrete. In others the wires are bonded to the concrete of the beam using grout made of a mixture of sand, cement, and water. Other systems cast the wires in the concrete and then tension them after the concrete has set—this is called post-tensioning.

Whatever technique is used, the central principle is always the same—to impart to the beam some initial stresses opposite to those that will be applied during its working life. In this way the net effect of the applied loads is reduced and weight is saved. A good example of its use is the modern London Bridge opened by Queen Elizabeth in 1973. It consists of three spans of prestressed concrete box girders built on the same location as Rennie's bridge mentioned in Chapter 2.

Just as the stories of Beauvais Cathedral and the Dee Bridge have helped us to see the importance of how structures might fail so another big learning event for bridge builders was the collapse of the Westgate Bridge while it was being erected over the River Yarra near Melbourne in Australia in October 1970. The full-length bridge was to be five spans of steel box girders, together with prestressed concrete approach viaducts, 8500 ft in total length and carrying two 55-ft-wide carriageways. Unfortunately there were site huts immediately beneath the span that collapsed and so in total 35 people were killed. The Milford Haven Bridge in Wales had collapsed a few months earlier, also during erection and also designed by the same engineers. As a consequence some strengthening work was being done on the Westgate Bridge. The Royal Commission of Inquiry report said that the immediate cause of the collapse was that a number of bolts had been removed from a cross splice in an upper flange steel plate near to mid-span.[11] They were removed to straighten out a buckle that had been caused by kentledge (heavy weights) that had been used to overcome some difficulties caused by errors in the camber of the bridge.

Construction work began in April 1968 but it soon became clear that labour problems were causing delays. Eventually a new contractor was appointed to erect the steel work but there was confusion, lack of cooperation, and antagonism between the various companies involved in the project. These difficulties were compounded by the novelty of the method chosen for erecting the steel boxes. The spans were assembled from prefabricated boxes 16 m in length into two halves on the ground. These halves were a full span length and half of the final width. Each half was jacked up into place on top of the piers, landed on a rolling beam, and moved into position. The two halves had to be bolted together up in the air. However, because the cross sections of the two half spans were unsymmetrical there was significant horizontal bowing of the half spans outwards.

When the first half span was lifted off its temporary staging on the ground the projecting flange plates had buckles as much as 380 mm deep. Because time was pressing it was decided to do any repairs in the air. Engineers on site checked the stresses but unfortunately made some basic errors. When the new contractors took over, the vertical difference in height of the two halves apart from the amplitude of the buckles was about 89 mm. They decided to undo some of the bolts in a cross splice to join the two halves and this worked for one span. So when they came to the next one they were reasonably confident they could handle it. They devised a better method using jacks to pull the two halves together but they were still not aligned properly so they loaded the higher north half span with large concrete blocks, each about

8 tons in weight, which just happened to be available. A major buckle developed but there they were able to jack the plates together. However, a diaphragm could only be partially connected. They removed the kentledge and undid some of the bolts to get the diaphragm fully connected. The buckle moved and one of the boxes was felt to settle. Some 50 minutes later the whole structure collapsed.

Any short account cannot describe the complexity of this incident. It is clear, however, that although technical reasons were the immediate trigger which caused collapse, the whole situation was brought about by a series of unfortunate coincidental circumstances. The structure itself was an advanced structural engineering design. The design engineers were acknowledged leaders in advanced bridge design and were working at the limits of existing knowledge. The deciding factors were the financial, industrial, professional, and scientific climate; the technically advanced design; and the novel erection procedure. The reasons for the failure were just as much those of human relationships and social organization as they were technological.

The man in charge of the structural design of the Westgate Bridge was Oleg Kerensky. Born in St Petersburg in 1905 into a well-to-do middle class family, he died in the UK in 1984 respected around the world as one of the leading bridge builders of his generation. Unfortunately two of his major bridges collapsed at the end of his career—causing him great distress.

Oleg's father was Alexander Kerensky, who in 1917 after the Revolution became Prime Minister of Russia until the Bolsheviks took over in the October Revolution. Oleg's childhood was happy, but the overthrow of Kerensky's government destroyed comfortable family life. Oleg escaped with his mother, Olga, and brother to London. Oleg spoke no English but by 1923 he and his brother had the necessary examinations to go to university. Both chose the Northampton Engineering College, London, then part of the University of London, now the City University, and both graduated in 1927. Oleg worked for a time as a young engineer on the design of the Sydney Harbour Bridge and hence met Sir Ralph Freeman (see Chapter 2). He then became a partner of Freeman Fox Consulting Engineers in 1956. He took responsibility for the design of the M2 and the Medway Bridge, which was at that time the longest concrete span in the world; the Erskine Bridge over the Clyde; and the Avonmouth Bridge, part of the M5 project. He led the design of the Severn, Milford Haven, Auckland, and Westgate Bridges as well as many smaller ones.

In his Presidential address to the Institution of Structural Engineers in London on 10 October 1970 Oleg said,

> As you all know, a few months ago a span of Milford Haven Bridge, with which I am intimately concerned, collapsed during construction, killing four men. All I want to say to the young engineers who may be listening to this Address or may find time to read it later, is that this everyday burden of responsibility, this ever present fear that 'there, but for the grace of God, go I'—is one of the crosses a professional engineer has to bear, be he a consultant or a contractor. The higher one rises the harder one falls, but if honest endeavour and vigilance are the basic cornerstones of one's efforts, no matter what disaster may occur, at least one's conscience can remain clear. ... I would like to finish with the motto of my life: 'Work to the utmost of your ability, accept success or failure with equal humility'.[12]

Those words were prophetic because only a few days later, on 15 October 1970, the Westgate Bridge collapsed, killing some of his own colleagues. Oleg's anguish can only be imagined. The Royal Commission subjected him and others to intense questioning. Oleg maintained until the end that the design was fully sound and that the fault lay with the contractors. Oleg Kerensky was a man who attracted attention simply through the strength of his personality. He retained a strong love for his native country, Russia, but never returned. Those who worked with him say that his sense of loyalty and need for teamwork was intense. He believed in hard work and had no sympathy with slackness of any sort. He had great integrity, real fairness, kindness, and humility. If he did have a fault, wrote Michael Horne in an obituary, it was that he trusted others too much.[13]

As a young engineer Kerensky had worked with Freeman on the Sydney Harbour Bridge with those enormous truss arches. His Westgate Bridge when completed was supported by cable stays. He knew that reading a bridge requires an appreciation of how arches, beams, trusses, and suspension structures can work together. In the next chapter we will look at trusses and how they are really a particular form of beam.

4

ALL TRUSSED UP
Interdependence Creates Emergence

Throughout history bridges have been used as mystic symbols. Perhaps the ultimate unknowable mystic journey is the passing from this world to the next. According to the Greeks you cross a river with the cantankerous ferryman Charon. In Norse mythology you walk over a rainbow arch bridge called Bifrost. The Chinvat Bridge between hell and paradise in Zoroastrian myth was made by Mazda (God). Al Sirat, also called the bridge of Jehennam, is a Muslim bridge to Paradise—only the good pass quickly, the wicked fall through into Hell. The Telumni Yokuts of North America must cross a long frail bridge over a stream that separates them from the land of the dead. The Semang of Malaysia must navigate a bridge called Balan Bacham to get to the magical island of Belet. The Ossetes, Armenians, and Georgians all have similar traditions. In ancient Rome, before Christianity, the pontifices (plural of pontifex) were the priests who bridged the gap between gods and men. The Pope is 'Pontifex maximus', which literally means bridge builder of the highest or greatest.

Many early bridges had chapels built on them. Travellers could pause and give thanks for a safe journey and pray for a safe arrival. Today only a few remain—for example, at Bradford on

111

Avon which later became a 'lock-up' (Figure 7). In the twelfth century the remarkable religious order of the Pontife brothers or Frères du Pont was formed in France. According to Frederick Robins they promoted the building of bridges but also established ferries and hospices on river banks near the crossings and collected tolls.[1] Their best known work is the St Bénézet Bridge finished in 1188 at Avignon.

From very early times bridges have been at the heart of the way we think about the human condition, the enigmatic connection between heaven and earth and between the natural and the supernatural. Bridges are links between the known and the unknown, they are places between places, they are part of how we try to deal with what we don't understand. As a consequence bridges stimulate new ideas and new understandings—often with interesting and colourful traits as the myths, ancient and modern, reveal.

Common sense tells us that actual physical bridges are much simpler since they are real, tangible products of human endeavour—they may be complicated but they are knowable. Builders of physical bridges only have to deal with facts, whereas mystical bridges are complex and unknowable. The distinction, however, is too easy. There are actually finite limits to the extent to which physical bridges can be known and their behaviour predicted and controlled. If we did have absolute understanding and total control then no bridges would ever fail—that is manifestly not the case. On the very rare occasions when we go beyond those limits and bridges collapse—a man-made disaster—we lose control, there is damage, and people are killed. For a while chaos ensues, but we begin to claw back control as we search for survivors in the wreckage and begin the cleaning-up process.

The cycle of events is just as a natural disaster such as an earthquake, major flooding, or tsunami—though it is only the big ones that hit the headlines. The limits of bridge behaviour that we have exceeded are boundaries of risk—a concept poorly understood (see Chapter 6). It is an unfortunate and uncomfortable feeling that despite all of the knowledge we now enjoy, there are still aspects of the behaviour of bridges that are unknowable. Experience tells us that so often there are unintended and unwanted consequences of what we do.

The difference between the obvious complexity of mystical bridges and the more obscure complexity of real bridges is, surprisingly, a matter of degree. Whether we are thinking and behaving spiritually or practically, what we do is based on what we believe. It depends on what we think we know as well as on our understanding of the likely consequences. The dependability of the understanding that bridge builders have about real bridges is clearly much greater than our individual ideas about life after death. That is because their collective understanding is based on testable ideas—science and a lot of practical experience of what works and what doesn't. That is not to imply in any way that spiritual faith is not genuine and real to the believer and an important basis for the way we live our lives.[2] However, belief derived from faith is not directly testable as scientific and technical knowledge must be. Unfortunately in practice scientific knowledge is not simply factual and is never total. This means that there are always real gaps between what we understand, what we do, and how things turn out in practice.

This largely unrecognized arcane difficulty of the complexity of bridge building becomes less obscure if we include, as we should, human beings in our definition of what constitutes a real bridge. Bridges may be physical objects but they are built by people for people. Physical bridges are embedded in human social systems and it is a major historical mistake to see them as isolated from their social and environmental context. Indeed the misconception is part of the wider historical mistake of failing to see that we humans are part of the natural world and not separate from it—a mistake that has led to environmental damage and the problems of climate change. Thomas Tredgold's famous nineteenth-century definition of civil engineering as 'The art of directing the great forces of nature for the *use and convenience of man*' (my italics) shows just how deeply the old idea that man is free to manipulate the natural world at will was embedded in our collective thinking. Most of us now recognize that whatever we do, including building bridges, must be sustainably in harmony with the natural world. That harmony depends totally on how we understand our social and environmental context. The story of the Dee Bridge in the previous chapter illustrated how wrong we can be about the way a physical bridge behaves and what can happen as a consequence. Although we have long forgotten the primitive beliefs behind the sacrifices of virgins to safeguard the foundations of the first London Bridge, the story demonstrates that what we humans do is based on how we understand the world. Accounts of bridge failures illustrate the complexity of the gaps between what we know, what we do, and why things go wrong.

Indeed in mid 2007, just as I was drafting this text, a 216-m-length truss bridge on the I-35 built in 1967 over the Mississippi in Minneapolis collapsed suddenly, killing 13, injuring 111, and creating localized mayhem as cars and trucks fell into the water. First reports blamed corrosion and fatigue. The bridge was inspected in 2005 and said to be structurally deficient but not dangerous. Since then it was inspected annually, indeed on the last occasion in May 2007. Nevertheless it still failed suddenly and unexpectedly. A report from the American Society of Civil Engineers said that, in 2003, 27% of the 590,750 bridges in the USA were structurally deficient or functionally obsolete.[3] Is this catastrophic event and this appalling situation, regarding the infrastructure of one of the world's most advanced nations, simply a question of negligence or the outcome from a complex social and political system in a country with many demands on its resources? It is surely too facile to call it simple negligence, though many do.

The I-35 bridge was a truss bridge. There are many of them in the USA in various states of repair. In this chapter we will look at the way truss bridges work. Whenever you see a bridge, or indeed any structure, made of lengths of timber or steel joined together to form triangles then you are looking at a truss. Timber trusses are commonly used for the roofs of houses and small buildings. Steel trusses may be used for large spans, for example, in airport terminals or sports stadia.

Trusses are clever, highly interconnected structures. They can be vulnerable or they can be resilient. Trusses are good examples of 'emergence'—the way complex systems grow out of interacting simpler systems, an idea at the heart of modern theories of complexity. So the vulnerability or resilience of a truss 'emerges' from the way the parts of the truss are connected. Arthur Koestler coined a very useful word to help understand this idea of emergence. He suggested that a 'holon' should refer to something, indeed anything, which is both a whole and a part. So, for example, you are a holon and so am I. You are a whole because you are an individual. You are a part because you belong to various social groups such as your family or the company for which you work. Wholes have parts—your parts are your subsystems such as your bones and muscle, nervous system, circulatory system, and immune system. Each whole has characteristics and properties that 'emerge' from the interaction of the parts—your ability to walk and talk, for example. None of your parts can walk or talk on their own. The net result is that you are more than the sum of your parts because you have emergent properties. You are the result of the active interaction of your parts with your environment in a process we call living. This same argument works at every level. Looking inwards, your structural subsystem of bones and muscle is also a holon with its own emergent properties such as your body size or muscular dexterity. Looking outwards your family is a holon with its own emergent properties such as happiness or closeness.

We have already begun to see parts and wholes in the story of the bridge book. At every level from letters to words, phrases, sentences, paragraphs, sections, and chapters, each is a part and a whole. At every level that part of a bridge is more than the sum of its parts. So the meaning of a word is more than the sum of its letters—it has characteristics that the letters do not have which result from the relationship and interaction between the letters. Likewise a sentence conveys a meaning that is much more than the sum of its words. It has characteristics that the words do not have which derive from the relationship between the words. The properties of an arch bridge result from the relations and interactions between its parts. The thrust line cannot exist in separate parts—it only exists as a result of the relationship and interaction between the voussoirs. The beauty of a beam bridge derives from the relationship between the parts of the bridge.

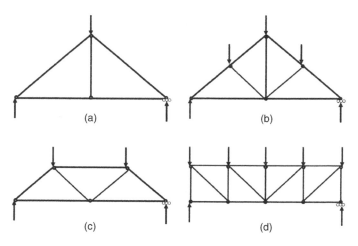

FIG 22. Some forms of truss

Just as beams evolved from the first clapper bridges the first truss bridges came from the inventive use of logs tied together. The first truss was probably made by erecting a vertical member at the centre of a beam and bracing it back to the ends of the beam or to the river bank. This is a simple 'king post truss' (Figure 22a). You may well have seen many examples in the roofs of old barns dotted around the countryside— they are part of the roof's frame shaped as an upside-down V, with a horizontal beam across the bottom at eaves level and a vertical timber from the middle of the tie to the apex of the rooftop. Other examples of king post trusses feature shorter diagonal timbers that run from the centre and rise to support the longer diagonals halfway down the roof (Figure 22b).

By Roman times the principles of the truss were being well used. One of the first was by Trajan's bridge builder, Appolodoros, who used diagonal bracing for a bridge over the Danube to build a trussed arch. Three parallel courses of timbers tied by braces or struts continued up to the floor of the bridge deck—shown on his column in Rome dated about 101 AD. The truss principle was used to build the temporary structures, or centring, necessary to build stone arch bridges for which the Romans are famous.

We can imagine what might have happened when there weren't any timbers long enough to span the whole way across the river. Two timbers would span to a central vertical post and then be pinned to each other and to the post. The pieces would be cut to fit together, holes pierced through them and pins or dowels inserted.

The small circles in the diagrams at the joints represent pinned or hinged joints. This is important because the members framing into a pinned joint are able to rotate freely about each other. There are no restraints against free rotation and so no internal bending moments. In other words a pin-jointed truss contains no internal bending moments because the joints have no stiffness.

In Figure 22 the small circle at the left-hand support of each diagram shows that the two members are pinned to each other and to the foundation. Therefore the foundation can resist any net force in the plane of the truss. Bridge builders usually resolve that force into its vertical and horizontal components as we shall see in a moment. The right-hand support is slightly more complicated. The way the diagram should be interpreted is that the two members of the truss are pinned to each other (the single pin shown) but together they sit on rollers. In this way horizontal forces are reacted internally in the structure but not transmitted to the ground because no reaction can develop—the rollers just move. This means that the external reaction from the ground is always vertical.

The net result is that there are potentially three external reactions at the supports resisting the applied loads. They are the two vertical reactions at each end and the one horizontal reaction at the left-hand support. The support conditions are therefore statically determinate because we can calculate them fairly straightforwardly. We do it by balancing the forces vertically; then horizontally, then with respect to the turning effect of the forces. This provides us with three mathematical equations that can be solved to find the three unknown reactions. In Figure 22, for simplicity, no horizontal forces but only vertical forces with two unknown vertical reactions are shown.

When we first discussed the king post truss of Figure 22a I glossed over an important detail which can now be addressed. I said that the horizontal member was a beam. However, when we introduced a pin joint at the centre of the truss the horizontal member was divided into two parts and therefore was no longer a beam. If we were to remove the vertical post the two horizontal members would just rotate about the pin at the centre and fall. In fact they can now only resist a direct force axially along their lengths—in tension or compression.

So what is going on here—haven't we just succeeded in weakening the structure by taking away the capacity of the bottom member to bend? In one sense the answer is yes. However, the bottom member can now be so very much lighter. As there is no bending, there is no need for a massive beam any more, and we can use something very light—a rope or cable perhaps.

We can see this by imagining that you are holding onto the truss by clinging onto the vertical post so that your weight is transferred totally to it. Because the horizontal members can no longer bend as beams, your weight, as a vertical downward force, must be transmitted up through the post to the top where it is resisted by the two inclined members. They transfer a compressive force down towards the river bank. As they are pinned to the ends of the horizontal member they push outwards at each end, so pulling it into tension. The forces all balance out in equilibrium—we'll demonstrate what is going on here using the triangle of forces to calculate the values in a moment. The bottom horizontal member is now simply a tie taking only tension and therefore from a structural point of view could just be a rope. Each of the elements or member of the truss is either being simply pulled or pushed along their entire lengths. The real magic of the truss is that together the struts and ties can resist many different kinds of loads and external forces from natural and man-made sources. The connected struts and ties interact, become interdependent, and have emergent properties.

Of course, as one might expect, there is always a problem. The inclined members are in compression and so they can buckle or break if they are too long. Soon therefore people began to think of different ways to use triangles to make bigger and better trusses over larger spans but with struts that aren't too long, as is evident with the two other common forms of truss shown in Figures 22c and 22d.

Now let's see how we can use the triangle of forces to calculate the internal forces in a truss. I have shown three different situations for the king post truss of Figure 23. Our strategy is exactly as for the tug of war rope in Chapter 1. There we made an imaginary cut in the rope. Now we will make three cuts around our joint so that we now have three internal forces acting at that joint which must balance. However, at each joint we need to be clear which are internal forces (shown with dotted arrow) and

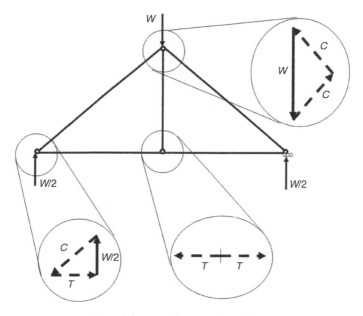

FIG 23. Internal forces at three cuts in a king post truss

which are external forces (shown with full arrow). So at the left-hand support there is one external force which is the vertical reaction $W/2$. We can therefore find the values of the internal forces C and T using the triangle of forces. C and T must act axially along the length of the members. Their directions determine the angles in the triangle. We can therefore draw the triangle to scale and measure off the values.

Next let's look at the centre joint. There can be no vertical force in the vertical post. You'll recall that an internal force will only be created when there is something to act as a reaction. At the bottom end of the vertical post, at the centre joint, there is no other member or external reaction to resist an internal force.

121

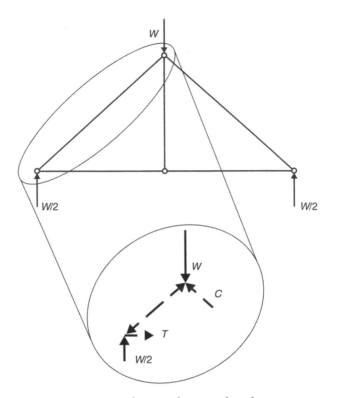

FIG 24. Internal tension force in rafter of truss

The horizontal force T is simply transmitted through the centre joint. The triangle of forces has flattened to a straight line. As a consequence there is no vertical force from the vertical post at the top joint. The vertical force W now balances the two internal forces C from the sloping rafters.

In Figure 24 you can see how the forces of Figure 23 create the internal compression C in the sloping rafters. At the bottom the external vertical reaction of $W/2$ and the horizontal internal force in the tie require the member to push back with the force C. At the apex the external force W and the internal force C in the other inclined rafter require the member to push back with a force C.

Bridge designers, of course must deal with situations where there are more than three members framing into a joint. However, no matter how many members there are we can only calculate two unknown internal forces by resolving the forces vertically and horizontally. If this is the case at every joint in the truss then the truss is statically determinate. Where this is not the case then the truss is termed statically indeterminate. The bridge builders must find other techniques using the energy stored in the bridge—as we'll see later in this chapter.

It is quite a long-winded process to draw a triangle of forces for each joint in a truss. So instead bridge builders usually resolve the forces into horizontal and vertical components. They replace a single force by two forces. They do this because where a number of forces in different directions come together at a joint they need to find the resultant, i.e. how they all combine together. This lengthy process can be shortened by finding all of the horizontal components and all of the vertical components which can then simply be added up. The result is a total horizontal force and a total vertical force—a pair of components—which represent the total effect of all of the forces combined. They can then be converted back into a single force if we wish.

When the joint is in equilibrium then these resulting component forces in the horizontal and vertical directions will be zero. We can use this knowledge then to find forces at the joint which are unknown—as long as only two are unknown. We form two equations—one by making the sum of the horizontal components equal to zero and one by making the sum of the vertical

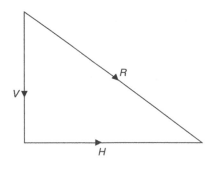

FIG 25. Triangle of forces

components to zero. We now have two equations and two unknown forces. The algebra of simultaneous equations allows us to solve them.

So how is a single force 'resolved' into its two components? Remember a force is a vector arrow of a certain length pointing in a certain direction (Figure 25). So what we need to do is make a right-angled triangle of forces. The three forces are in equilibrium. It follows therefore that if the directions of any two forces are reversed they then become equivalent to the third force.

If that is unclear think of it this way. A force is the distance from one vertex of the triangle to another. You can get from one to the other in two different ways—the most direct way—the single force—or the indirect way by the other two forces. In this way the force on the longer side (the hypotenuse) can be replaced by the forces on the other two sides with their directions reversed. For example, a force of 5 N can be replaced by forces of 3 N and 4 N in a right-angled triangle.[4]

This technique of resolving forces is used to calculate the internal forces in a truss. In a moment we'll start to look at how truss bridges have developed over the centuries but before that, let's just pause to look again at how ties (in tension) and struts (in compression) are strong because there are some important details we have neglected. In Chapter 1 we examined the strength of a tug of war rope and said that the fibres must resist the internal force set up when it was pulled. We noted the interesting fact that the strength of a tie—a member in tension—does not depend on the length of the rope but only on the cross-sectional area of the tie and the material from which it is made. It is therefore relatively simple to find out the tensile strength of a tie—we just decide on a material and find the force that makes a unit area of the material fail as we found out in Chapter 1. The strongest material in tension is steel—particularly in the form of wire rope. Other common construction materials like concrete and glass or copper and aluminium are very weak in tension.

Of course nothing is ever quite as simple as it seems. If we stretch a piece of steel in a testing machine we can plot a graph of how the stress changes with the strain. If we find that all of the points lie on a straight line as we load and unload the piece then we say that the material is linear elastic. However, at a certain high stress the behaviour changes—sometimes very abruptly—and the points no longer follow the straight line. Indeed the strain suddenly gets very large. This change point is called the elastic limit and the material is now said to be plastic because the rate of increase in strain is much larger. If we continue to apply load to the steel the stress will continue to increase to a maximum—the ultimate tensile stress—the strain will get even bigger until the stress begins to fall off no matter how hard we try to apply more load. Eventually the steel will rupture—the rupture stress.

Inevitably some materials do not behave in this way. In particular many, such as aluminium, do not show a distinct elastic limit.[5] More importantly some materials, like cast iron, are brittle and break very suddenly with very little strain and hardly any ductility. Many structures are designed particularly to exploit the strength and ductility of materials in tension. We will examine some of them in the next chapter on hanging structures.

In Chapter 1 we saw that the strength of a structural member in compression depends on a number of complex factors. Very short members just squash in a process which is more or less a reverse of pulling apart in tension. The determining factors are the material, the elastic limit, and the ultimate squash load when the material can no longer carry any load at all.

Longer struts are more of a problem, however, because they bow out and buckle. A very slender strut buckles very easily, especially when not perfectly straight. On the other hand a strut is stronger if the ends aren't allowed to rotate. The standard end condition, against which others are compared, is a pin at both ends of the strut so that they can rotate freely. This is how most modern trusses are designed to behave. A flagpole or a lamp post are examples where one end is held rigidly (in the ground) and the other end (the flag or lamp) is entirely free. The vertical leg of a tower crane that you typically see on a construction site can carry a significant compression load and is free to move at the top but constrained from rotating at the bottom.

The slenderness of a strut not only depends on its length and initial straightness but also on the shape of its cross section. There are two main forms of slenderness that are important—longitudinal and torsional. Longitudinal slenderness is the extent to which a strut is long and thin. Torsional slenderness is its susceptibility to twisting. When a strut buckles, it takes the line of least resistance—it finds the easiest way to move out of the way of the squeezing load, either by moving sideways or by twisting. The best shape for the cross section of a strut, therefore, is a circular tube because the distribution of material is equal in every direction. This is so for both longitudinal and torsional buckling. A deep I-beam makes the worst kind of strut because although it is strong in one direction it is very weak in the other.

As we noted earlier the Romans used the truss to build their arches. Trusses were used after them but there was little development. By the sixteenth century Palladio was only able to write about four different truss designs in his *Four Books of Architecture*. These included a multiple king post truss and arches of interlocking triangles.

Then there was a big surge. It happened mainly in North America with the building of the railways. It was fed by the plentiful supply of timber—ideal for trusses. You may have seen some examples of the early, rather rickety looking American timber trestle railway bridges in films or on TV. Timber-covered truss bridges were built too. The first one that I ever saw looked like an isolated, large wooden shed sitting over a stream somewhere in New England—somewhat incongruous. Covered bridges are more than just simple sheds, however—the structure is usually a large timber truss and robustly well built. As prominent landmarks, dating mainly from the nineteenth century, they are of significant interest to historians.

The tradition of covered bridges did not start in the USA. There are many in central Europe. The 204-m long Kapellbrucke (Chapel Bridge) at Lucerne was built in 1333 with paintings depicting historical events in the town. Unfortunately it was badly damaged by fire in 1993. There are also covered bridges in Asia—for example, at Lhasa in Tibet. The seventeenth-century 'turquoise-crowned' bridge was unfortunately torn down in 1993. In China some bridges, such as those in Guizhou, are called 'wind and rain bridges'. Some famous stone-covered bridges include the Rialto Bridge in Venice, Italy, and the Bridges of Sighs in Venice, Cambridge, and Oxford. Modern covered bridges are usually for pedestrians.

The wooden covered bridges of the USA are perhaps the most curious. Why are there so many and why were they covered? Early wooden bridges tended to deteriorate rapidly. People found that when the structure was covered, and hence protected, its useful life was substantially lengthened. As Ed Barna, writing about the bridges in Vermont, USA, says they were 'one of the finest expressions of America's golden age of woodworking'.[6] He says that a contractor could tell whether a man seeking adze work was truly an experienced craftsman by whether he still had all of his toes![7] The small state of Vermont has the greatest density of wooden covered bridges in the world with 107, including those shared with New Hampshire. Oregon has 52; there are 94 in Quebec, and 65 in New Brunswick.

The Hartland Bridge opened in 1901 and crossing the Saint John River in New Brunswick is, in 2009, the longest covered bridge in the world at 390 m. The Blenheim Bridge in New York State over Schoharie Creek has the biggest single span of 70.7 m and was built in 1855. The Smolen-Gulf Bridge over the Ashtabula River, Ohio is 182.9 m long and was opened in 2008.[8]

Covered bridges were sometimes nicknamed 'kissing bridges', as they provided a place to meet in seclusion. They had many other uses too—they served as bulletin boards for local events with advertisements for such things as Kendall's Spavin Cure for equine ills. They hosted meetings and served as children's clubhouses and playgrounds, and people fished from them. There was also a dark side, unfortunately, since robberies on these bridges at night were not unknown.

Covered bridges have inspired many people and these included novelists. *The Bridges of Madison County* is a novel by Robert James Waller, later made into a Hollywood motion picture. It's 1965 and Robert (Clint Eastwood) has come to Iowa to photograph the covered bridges. He can't find the Rosamunde Bridge, which in reality is the Roseman or Oak Grove Bridge of 32.3 m span built in 1883. He meets Francesca (Meryl Streep), who is a lonely housewife, and she helps him and they fall in love.

Many of the trestle railway bridges in the USA were short-lived and nearly all were trusses of one form or another. Many different variations of the basic triangulated truss form were built. Patents were taken out by, among others, Timothy Palmer, Lewis Wernwag, Theodore Burr, Ithiel Town, and Squire Whipple for timber and metal trusses. Other well-known truss forms were the Howe, Warren (see Figure 22c), Pratt (see Figure 22d), Bollman, and Fink.

This surge in the development of truss bridge building was all based on principles that we have looked at in some detail for the king post truss. However, there are two further important details we need to consider. First, just what is the difference between a truss and a beam? They seem to be doing the same job and yet we know from Chapter 3 that beams carry bending moments. A truss with pin joints has only struts and ties—where are the bending moments? Second, how would the behaviour of the truss change if the joints weren't pinned but fixed?

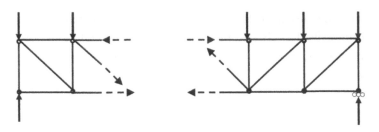

FIG 26. Internal forces in a Pratt truss

We'll use the Pratt truss of Figure 22d to help answer the first point. We saw earlier, for the king post truss, how to calculate the external force reactions if we know the external applied loads—so we can do exactly the same for this truss. Now let's look at Figure 26 where an imagined cut has exposed some of the internal forces, depicted as the three dotted arrows.

Now compare this diagram with Figure 18c. The horizontal axial internal forces in the top and bottom members (often called chords) are resisting the turning effect just as the distributed stresses did within the beam. The internal tensile force in the diagonal member has a vertical and horizontal component which is resisting the vertical and horizontal shear.

In summary and in answer to the first point, the truss and the beam are balancing the same external forces but with different distributions of internal forces. They are just two ways of doing the same thing.

We can answer the second point—what would happen if the joints were fixed not pinned?—by recalling what happens when you restrain a degree of freedom. We know that the members of a truss want to rotate relative to each other. If we prevent that rotation by fixing the joint then internal bending moments will be created. In effect the truss becomes a connected set of beams.

Let's turn now to that great railway bridge that we began to look at in Chapter 1, the Forth Bridge in Scotland (Figure 4). We started to read its forces but now we can dig a bit deeper and begin to look at the way it works in more detail. Before we do, it's worth noting that some critics have accused Baker and Fowler of over-design—of being too cautious. We cannot know Baker's innermost feelings about the enormity of the task that faced him. Thomas Bouch's Tay Bridge had recently collapsed in 1879 and as a consequence Bouch had been sacked from working on the design of the Forth Bridge. Baker said in a lecture:

> The merit of the design, if any, will be found, not in the novelty of the principles underlying it, but in the resolute application of well-tested mechanical laws and experimental results to the somewhat difficult problem offered by the construction of so large a bridge across so exposed an estuary as the Firth of Forth. ... If I were to pretend that the designing and building of the Forth Bridge was not a source of present and future anxiety to all concerned, no engineer of experience would believe me. Where no precedent exists the successful engineer is he who makes the fewest mistakes.[9]

Baker knew that success lay in being careful and thorough—a point that Bouch had not grasped. The Court of Inquiry into the Tay Bridge disaster found Bouch culpable and he died with a broken spirit the following year in 1880.

You will recall that we said in Chapter 1 that the top arms of the trusses of the Forth Bridge act at an angle to take a pull and the lower arms of the truss carry the thrust to the foot of the cantilever tower and the two sides of the towers balance each other. The downward forces from the towers sit directly on and compress the foundations. However, we looked at the Forth Bridge only at chapter level. Now in a more detailed look, the paragraphs are the trusses; the sentences are the struts and ties; the words are plates, bars, tees, angles, rivets, etc.; and the letters are the steel materials.

At first sight the bridge looks dauntingly complicated but we can begin to unravel what is going on. In order to read the bridge the first thing to note again is the importance of the triangle. The towers, cantilevers, and suspended spans are all triangular trusses. Every member in the bridge is therefore either a tie or a strut—in direct tension or compression.[10] What is more you can identify each one because all of the struts are solid tubes and all of the ties are themselves latticed trusses. So if you see a solid member it's a strut. If you see a latticed member it's a tie.

Now look at the towers. Each one has four vertical struts at the corners which lean in slightly. Then looking from the side there are two struts as cross bracing that form triangles. In fact under some loading conditions they are in tension but the critical internal force is compression, which is why they are tubes. Next look at the cantilevers—each has six bays, each with a strut and a tie forming the cross bracing. Finally the suspended span is a simple triangulated truss.

At every joint the forces must balance based on our triangle of forces. If you cut a cross section through a strut then you would see steel plates curved and riveted together to make tubes. Remember that the strength of a strut depends on its length, the way the ends are held, but most of all on the way the material is distributed in the cross section. A closed tube makes the best strut because the material in the cross section is distributed uniformly—no one direction is weakest—there is a uniform resistance to buckling in all lateral directions. All struts in the Forth Bridge are tubes and many are circular but some have flattened sides so that the connections can be more easily made. Each strut has a number of diaphragms internally to help the tube keep its shape.

The strength of a tie depends simply on the cross-sectional area of steel available. On the Forth Bridge the tie forces are large so the cross section is built up from tees, bars, and plates. If you cut a cross section through a tie then there are four main booms, which are braced on all four sides.

The wind blows strongly up the Firth of Forth and so there can be some large wind pressures on the bridge. So as well as the trusses to carry the self-weight of the bridge and the weight of the trains along the length of the bridge, some trusses are needed across the width of the bridge to resist the wind pressures. Figure 27 shows a tower of the bridge during construction looking along the length of the bridge. The legs as struts are clear as are the cross diagonals between the left-hand pair of legs. The lattice tension members growing up from the bottom will form, with the legs, trusses across the width of the bridge to resist sideways or lateral wind loads. Each one is therefore always in tension.

The bridge is a massive steel structure and will therefore expand and contract with changes in temperature. There are certain fixed points, and movements are absorbed by sliding bedplates, and rocking posts at one end of each central girder, and by roller bearings in the two cantilever end piers.

FIG 27. Erection of a leg of the Forth Railway Bridge

The final chapter is, as always, the foundations. The forces at the foot of the towers from the four columns are transferred through bedding plates to a circular granite pier rising 18 ft above high water from permanent caissons. These were large watertight chambers in which the men could excavate in the dry until they were well founded. In the picture (Figure 28) the upper part of the caisson was temporary. The lower part became the actual foundation of 21.3 m diameter with varying heights and topped by a 7.3-m-high tapered section. The airtight floor at the bottom was supported by four strong lattice girders 5.5 m in height. You can see the cutting edge at the bottom where the men were working. The caisson was eventually filled with concrete and rubble.

Unfortunately one of the caissons accidentally tilted and in rough tides and sank unevenly into the mud. It took months to refloat it and sink it in the right place.

Wilhelm Westhofen, a German engineer, worked on the bridge and in 1890 wrote a very detailed account of its design and construction:

> ... by far the best view of the bridge is obtained from the river, whether above or below, at a distance of a mile or so, the structure rearing itself to a great height, and being backed only by the sky. Thus viewed, its simple lines, its well-proportioned parts, its impressive air of strength and solidity and yet of lightness and grace, never fail to strike the mind of the beholder. Four-square to the wind and immovable it stands![11]

The Forth Bridge is an icon for Edinburgh and for Scotland. Middlesbrough has a less well-known icon—its transporter bridge, another magnificent truss structure.

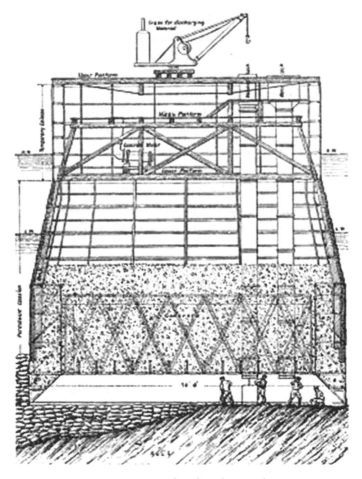

FIG 28. Caisson of Forth Railway Bridge

Transporter bridges are air-borne ferries. Pedestrians and vehicles are carried over a river in a gondola or platform structure slung from under the bridge. The bridge itself is usually quite high up to allow the tallest ships to pass underneath. The gondola is hung from a trolley which runs on rails along the bridge. It operates at the height of the road so that cars and pedestrians can drive or walk straight on. Of course it takes time to load and unload the vehicles.

Transporter bridges are rare and vary in structural form—some are truss bridges and some are suspension bridges. As traffic volumes increased they became obsolete.

There are two working transporter bridges in the UK and several in other parts of the world including USA, Germany, France, and Spain. The Middlesbrough Transporter truss bridge was opened in 1911 and is 49 m tall. The Newport, Gwent suspension bridge, opened in 1906, stands at 75 m. The Bank Quay truss bridge in Warrington over the River Mersey was opened in 1915 but was closed in 1964. It is now an English Heritage listed building. A gondola is proposed for the Royal Victoria Dock Bridge in London and will be a glass passenger cabin.

The Middlesbrough bridge operates a regular 15-minute service and can carry up to 200 people and nine cars in just over a couple of minutes between Middlesbrough and Port Clarence. It is almost 260 m long and structurally is reminiscent of the Forth railway bridge in that it is two cantilevers but without a suspended span in the middle. From a distance it looks almost like two tower cranes touching at centre span. In fact it is two almost independent vertical truss towers supporting trusses which cantilever back by nearly 43 m and cantilever forwards across the river nearly 87 m, where they meet and mutually support each other. The anchor cantilever is held down by cables as back stays.

Transporter bridges are massive and fixed in place. In some situations there is a need for the very opposite. Sometimes bridges that can be moved around flexibly and made available quickly for use in many different situations are required.

Donald Bailey was a civil engineer in the British War Office during World War II who turned his mind to this problem as part of the war effort. His idea was to provide temporary bridges that could be built quickly, under difficult conditions. He knew that military bridges had to carry heavy tanks and he envisaged spans of up to 60 m. He wanted standardized and interchangeable components, light enough to be handled by up to six men and transported on a 3-tonne truck. The equipment had to be highly adaptable.

Donald Bailey was born in Rotherham in 1901. He went to Rotherham Grammar School and the Leys School, Cambridge before graduating as a civil engineer from the University of Sheffield. For a time he worked for London Midland and Scottish railway until he joined the War Office in 1928 as a civil engineer at the Military Engineering Experimental Establishment in Christchurch. In 1940 he presented his ideas at a conference on the problems of providing temporary spans capable of taking heavy loading. They were immediately approved. The significance of his final contribution was immense. General Montgomery is reported to have said that without Bailey Bridges the Allies would not have won the war.

A Bailey bridge has four chapters—the panels, the decking, the bearings, and the fixings, which are used to connect all of the 'bits'. The panels are prefabricated welded steel cross-braced rectangles 3 m long. They can be stacked side by side or on top of each other and pinned together to make trusses as needed to cover the required span. The deck consists of steel I-section cross beams or transoms clamped onto the panels to hold them together. Smaller beams or stringers span between the transoms. Finally the deck is surfaced with timber or steel panels.

Each panel unit built in this way forms a single 3-m section of bridge with a 4-m wide roadbed. After it is complete it is typically pushed forward over rollers on the bridgehead, and another section built behind it. The two are then connected together with pins at the corners of the panels.

Most bridges can be assembled in a matter of days by a small crew using common hand-tools. Connections are pinned, bolted, or clamped with no welding—disassembly is also straightforward. The components can be put together in a number of configurations to accommodate a range of span and capacity requirements; e.g. panels can be connected together in two or three storeys to make the bridge stronger and capable of carrying heavier loads.

The essence of Bailey's design was its simplicity, which enabled mass production. Some 700,000 panels were made—about 350 miles of bridging. Its versatility was legendary. After the war many Bailey bridges were used while the shattered infrastructure was rebuilt. As a child in Derbyshire I well remember driving over a Bailey bridge that was not finally replaced until well into the 1950s.

Donald Bailey was awarded an OBE in 1944 and knighted in 1946. He died in 1985 in Bournemouth.

Some of the key components in Bailey bridges are the bolted, pinned, and clamped connections. Connections may be small details but they are crucial for the integrity of trusses—they are good examples of the well-known aphorism 'the devil is in the detail'. Without good connections bridges will fail.

Bolts are long cylindrical bars with a head at one end and threaded at the other to receive a nut. Bolts are generally hand-tightened using a spanner. However, special high-strength friction grip bolts are tightened with a torque wrench to clamp the plates together so tightly that the friction between the plates resists the shear force applied to them. Timbers are often connected using dowels—they were used extensively in covered bridges. Dowels are wooden headless pegs driven into holes through the timbers. They were perhaps the forerunner of rivets, which are metal rods or pins with heads usually made of steel. The rivet may be cold but in big bridges such as the Forth Railway Bridge they were red hot. They were inserted in a hole drilled through the pieces of steel to be joined. The head is held tightly while the other end is hammered to form another head and the metal pieces are gripped. The use of rivets died out after World War II and they are hardly ever used now.

Welding is used on almost all modern steel bridges. It is a process in which the pieces of metal to be joined are heated to a high temperature. Molten filler is placed carefully so that it fuses or coalesces into the metal to create the join. The two main types of weld are fillet (see Figure 29a) and butt (see Figure 29b). Fillet welds are used when the plates are at an angle—usually 90°. Butt welds are suitable for joining plates in line—for example, the plates of a box girder or the two ends of a tube.

As always there is a downside. Welding does to a considerable degree depend on the skill of the welder. Automatic welding machines are used but must be set up very carefully. The welder

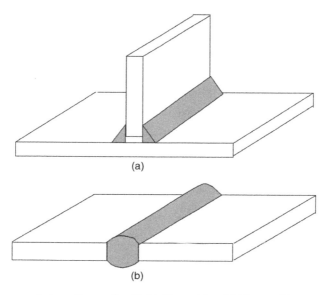

FIG 29. Types of weld: (a) fillet weld and (b) butt weld

must be watchful that the metal around the weld does not become brittle due to metallurgical changes in the area affected by the heat. He must make sure that the weld does not have significant defects such as extraneous material getting into the weld (this is called slag inclusion). He must avoid the weld becoming porous or the edges being undercut. He must ensure that there is good penetration and that the metals fuse properly. He knows that the putting in of so much heat may cause residual stresses that can sometimes cause problems. Residual stresses are complex self-balancing, locked-in internal stresses retained after a piece of material has been strained from either heat or external force. Despite these risks, when done properly under controlled conditions, welding is a highly successful form of connecting the elements of modern steel bridges.

All bridges must incorporate expansion joints. They must be designed to accommodate the movements of a bridge due to temperature changes. Bridge bearings are the supports to a bridge on a bridge pier that carry the reaction forces to the foundations and control the movement of the bridge. The main types are metal rockers, rollers or slides, or laminated rubber and steel. Rollers are literally round hard steel bars that the bridge sits on to allow limited horizontal movement. A rocker bearing is a device that is free to rotate but not to move horizontally unless it is carried on rollers. Many modern bearings are elastomeric laminates. These are layers of rubber with steel plates to allow some rotation and some horizontal movement.

The importance of connections and bearings is illustrated vividly by the story of the Second Narrows Bridge. In June 1958, two spans of a six-lane highway bridge under construction across the Burrard Inlet in the Greater Vancouver area, Canada, collapsed suddenly, killing 18 people. The story also highlights the possible personal consequences of an error in design. Among those killed were the young designer of part of the falsework and his immediate boss. He made a simple mistake in his calculations and his boss failed to spot it. He calculated the shear stress on a steel beam in a supporting grillage using the whole cross-sectional area of the beam, rather than the area just of the web. The value calculated of 41.4 N/mm^2 was half the value of what should have been obtained and so he decided that stiffeners were not required.

As well as killing people and creating chaos for a while, the collapse would have released an enormous amount of energy. In

Chapter 1 we wondered about the energy released if the rope in the tug of war broke. It would be noisy as the fibres split and separated and the two teams would have fallen over backwards as the resistance to their pull suddenly disappeared. So where does this sudden release of energy come from?

As the rope was pulled by the tug of war teams the internal tension force caused it to stretch so that every little element of fibre under tensile stress has strained. The product of the force (stress) and the movement (strain) is work. Energy is capacity to do work. The pulling of the teams on the rope locks in a type of energy called strain energy—energy due to the strains within the rope. Other more familiar forms of energy are potential energy and kinetic energy. A heavy book on a high shelf has potential energy because if it is knocked off the shelf it will fall to the floor—before it falls it has the potential to move and hence do work. Kinetic energy is that contained in a moving body so when you are driving in your car you have kinetic energy. Everything that applies to a rope also applies to a bridge—indeed to any solid body deformed by applying loads. So strain energy is the potential energy in a deformed body, i.e. one that has internal strains and stresses.

It turns out that the total amount of strain energy in a bridge is of interest to bridge builders because they can use it to calculate internal forces in situations where they would just be stumped otherwise. Indeed strain energy is the basis for all modern computational techniques for the analysis of the forces in bridges. Without an understanding of strain energy no modern bridge could be built.

In Chapter 3 we found we could calculate the reactions of a simply supported beam by the use of simple equilibrium. The vertical forces must balance, the horizontal forces must balance, and there must be no net turning effect if the beam is to remain stationary and not fly off in some arbitrary direction. So, for example, in Figure 19a if we balance the vertical forces then $W = R + R$ and so $R = W/2$. If there is to be no net turning effect about any chosen point, say the left-hand end support, then R times the span length must be balanced by W times half the span length. Hence again $R = W/2$ and there are no horizontal forces. As I said earlier problems like these are called statically determinate because we have at maximum three equations with three unknowns and we can easily find all of the internal forces in the bridge.

In Chapter 3 we glossed over how we might determine the reactions at the ends and in the middle of a continuous beam of two or more spans. Unfortunately there are too many unknown reactions and we cannot solve their values simply. These types of bridge are statically indeterminate and present us with a problem.

Fortunately in 1873 Albert Castigliano in Italy came to our rescue. He developed two mathematical theorems based on strain energy that can be used to analyse statically indeterminate structures. However, his theorems only applied to linear elastic materials as defined in Chapter 1.[12]

The principle of least work says that for a bridge where the support reactions do not move, the actual distribution of internal forces in the bridge occurs when the strain energy stored is a minimum. As the bridge responds to the loads applied to it, all of the internal force and movements, the stresses and strains settle down to values such that the total strain energy in the bridge is the smallest it can possibly be. In that settling-down process the bridge does the least work it can possibly do, given all of the constraints on the way it can move.

The strain energy of a bridge became a central concern of the bridge builder, for it was a way to determine the internal forces that had previously not been available. However, it was still restricted to linear elastic structures. Eventually the least work principle was generalized in a way that can be used for modern bridges. The principle of the minimum of the total potential and the total complementary potential are used in finite element analysis using computers. The basic ideas are the same although the derivations are more 'mathematically complicated'.[13]

In brief they rely on the idea of a thought experiment. Virtual work is the work done when a known force is moved a known virtual displacement compatible with the constraints. The virtual (imaginary) displacements may or may not be the real ones.[14]

These new theoretical ideas are the basis for the design of all of our modern bridges—especially the architecturally adventurous suspension structures that we will look at in the next chapter. But before we do that let's look at how modern bridge builders are using these latest analytical techniques to design bridge trusses. One is an example of a novel structural form and the other of a modern form of timber.

The Liffey Bridge, Dublin, connects Lower Ormond Quay on the north side of the River Liffey with Wellington Quay on the south. The river is 51 m wide, although the central span of the bridge is 41 m between abutments. The truss is an asymmetrical parabolic arch truss but connected to concrete abutments in a way that transfers a bending moment to create a portal frame. As a consequence the forces in the truss are smaller and the members are slender, thus creating a sense of lightness and transparency. The deck is of slotted aluminium, supported off a series of secondary ribs running between the cross members. These are integral with the top booms of the truss and continue upwards to provide supports for the balustrade and aluminium bronze leaning rail. The bridge was designed so that most of the construction work could be carried out from the riverside of the quay walls. From temporary working platforms, a piling rig drilled rock anchors into the bedrock below the river silt to provide the foundations for the bridge abutments. The abutments were constructed without cofferdams, which instead required the contractor to work around the clock to take advantage of low tides. After the concrete pad foundations below the low-water mark were built, curved shells to smooth the flow of water were installed. Two prefabricated concrete fin haunches for each abutment were lowered into place in readiness to receive the metal bridge truss. The truss was fabricated off-site as a single piece and lifted into position onto the new abutments in November 1999 using a single crane.

The Flisa Bridge across the River Glomma in Norway is one of the biggest timber bridges in the world. In Norway, as in many countries, timber was the most common bridge-building material until the late nineteenth century when steel and concrete took over. Consequently very few timber bridges were built in the early twentieth century. However, a new product called glue-laminated timber (named glulam) began to be used around the 1960s. Glulam is made in manufactured lengths by gluing together layers of flat timbers which have been graded for strength. The flat timbers are joined end to end but the joints are staggered at intervals through the layers so no joints 'line up' through the thickness.

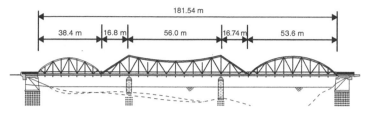

FIG 30. Flisa Bridge, Norway

These manufactured lengths are then cut and made into beams and other structural shapes for use in structures of various kinds. Glulam is a highly engineered product but unfortunately in the early days it gained a poor reputation due to bad detailing.[15] Inadequate protection from the weather in some cases and poor repair work resulted in a shoddy appearance. However, new research has demonstrated the potential of glulam.

The bridge has three spans of glulam trusses as shown in Figure 30. In cross section the two 'through' trusses sit side by side, 10.8 m apart to allow two traffic lanes and a pavement. The centre trusses are 90 m long but sit across a span of 56 m and so overhang at each end by 16.8 m. They have upper chord members which curve down from the piers where the trusses are at their deepest (9.5 m) to the centre of the span. The two end sets of trusses have arched upper chords and are simply supported by the overhanging cantilever of the centre truss at one end and the abutment at the riverbank at the other. One is 38.4 m long—the other is 53.6 m so that the total end spans are 55.2 and 70.4 m, including the 16.8-m cantilevers from the centre truss.

The top chords of all three trusses are braced together horizontally—effectively making three more trusses—but lying on their sides to resist the sideways force of the wind. The timber members were connected using steel plates embedded in sawn slots in the glulam. Dowels were jammed into holes drilled through the plates.

Liffey and Flisa are just two examples of the seemingly infinite variety of ways in which struts and ties can be joined together to make trusses. Trusses illustrate how the idea of emergence at many different levels is central to a reading of bridges. Even the obscure complexity of the limits of our ability to control events in the chaos of the very rare instances of collapse gives them a kind of mystical complexity through our difficulties in understanding risk. Also there are two important emergent properties related to risk that I mentioned right at the beginning of the chapter—vulnerability and resilience—that we have not covered at all—we will return to them in Chapter 6. Now we will turn to the exciting world of hanging bridges.

5

LET IT ALL HANG DOWN
Structuring Using Tension

Time travel fascinates many. We watch uneasily as the Daleks try to exterminate Doctor Who. We are relieved when he escapes by travelling backwards and forwards through time in his Tardis. We are engrossed as Arthur Dent and his friend Ford Prefect from Betelgeuse survive the destruction of the Earth to make way for a hyperspace bypass in *The Hitch Hiker's Guide to the Galaxy*. We are amused by Michael J. Fox as Marty McFly in the film *Back to the Future* as the only kid to get into trouble before he was born.

But all of that is in our collective fictional imagination. What do scientists say about time? It seems that many of the equations of theoretical physics, at the microscopic level, are symmetrical in time. This allows some to speculate that there may really be other possible worlds.

Nevertheless our everyday experience is unequivocal. We are born, we live, and we die—the arrow of time marches on in a one-way life cycle. We remember the past—not the future. Everything that we know of exists and changes through time. At the extremes some processes are very slow like the evolutionary and geological changes that take millions of years. Others, like changes in subatomic particles, are so fast that we need special equipment to detect them.

In this chapter we will look at the most exciting, absorbing, and interesting form of bridges—hanging bridges. They are exciting because they can be light and daring with long spans. They are absorbing because they can be refined works of art. They are interesting because at one level the flow of forces in hanging bridges is easy to understand; yet, when we dig down to a deeper level we find that they are statically indeterminate and are a blend of many time-dependent complex processes.

Bridges, like all man-made objects, have a life cycle. They are conceived, promoted, financed, designed, built, used, and dismantled. Time is particularly important for the builders of hanging bridges. In common with all other types of bridge some processes are relatively slow, like the deterioration of concrete that has been known to fall off in lumps from a neglected bridge. But the wobbly London Millennium Bridge of Chapter 1 illustrated how the slenderness and flexibility of hanging bridges makes them susceptible to some relatively fast processes. Other examples include the flutter of a tension cable, or indeed of a whole bridge, in the wind. A landmark event was the failure of Tacoma Narrows Bridge (see later in this chapter), which shook itself to bits in 1940 in a 40-mph wind.

So the theme of this chapter is change through time—change which may be slow or may be fast. By their very nature bridges enable change because through them people cross barriers and see a different world on the other side. Probably one of the earliest ways to cross a small stream or river, in dense foliage, was to swing from vegetation that happened to be handy—just like Tarzan. Eventually people probably made tight ropes to haul

FIG 31. Snowy Creek Bridge, New Zealand

themselves hand over hand across the gap. Soon they learned to make simple rope bridges which sway and wobble but are light and easily made and can cover large gaps. The first iron hanging bridges may have been in India and China—Harry Hopkins refers to one built by Emperor Ming in 65 AD.[1] Despite the fuss made about the wobbly London Millennium Bridge we still have simple wire rope bridges even today—see Figure 31.

Two perhaps more well-known and spectacular wobbly bridges that challenge the tourist are the Capilano Bridge in Vancouver and the Carrick-a-Rede Rope Bridge in Northern Ireland. Capilano was built in 1889 and rebuilt in 1956. With a span of 140 m and a height of about 70 m above the Capilano River it provides magnificent views of the Capilano Canyon and the forest. The current Carrick-a-Rede (rock in the road) rope bridge was built by the National Trust on the North Antrim coast in 2000. It crosses a chasm 24 m deep and 18 m wide which separates the north coast mainland from the very small Carrick Island. Apparently some visitors unable to face the scary walk back across the bridge have had to be taken off the island by boat.

Wobbly bridges may be acceptable in remote places or as tourist attractions but they are of little use as major highway bridges. The breakthrough for hanging bridges came when Judge James Finley built one with a level stiff bridge deck. His Jacob's Creek Bridge, Pennsylvania, built in 1796 or 1801 had wrought-iron chains instead of rope and a level deck 3.8 m wide slung underneath, over a 21-m span.[2] Finley was born in Maryland in 1762, was a member of the Pennsylvania House of Representatives and Senate, and was a judge for Fayette County. He died in 1828. Unfortunately his first bridge was demolished in 1833, several others collapsed, and none of his many other bridges have survived.

Finley patented his system in 1808 and published it in New York in 1810. The ideas spread quickly in the USA and UK, opening up all sorts of opportunities with a rapid increase in maximum spans. As a direct consequence of Finley's initial concept we have now developed and built many beautiful iconic structures such as Brooklyn, Golden Gate (USA), Clifton, Humber (UK), Normandie, Millau (France), Akashi Kaikyo, Tatara (Japan), and Tsing Ma, Stonecutters (Hong Kong). These bridges need careful nurturing and attention. They can still wobble and vibrate, especially in the wind or during an earthquake, but the level of excitation can generally be controlled.

There are basically two types of hanging bridges—suspension bridges and cable-stayed bridges. Hanging bridges often have a very high profile either because they are in a prominent position or are built to celebrate special events such as the millennium, the Olympic Games, or international exhibitions such as an Expo. That is why bridges can become *icons* of cities and regions and can have a high emotional impact. Hanging bridges also provide a great opportunity for *public art*—but more of that later.

The chapters of the book of a hanging bridge are the suspension system, towers, stiffening girder, bridge deck, and the foundations including bearings and anchors. Since we have largely covered the principles for the towers, stiffening girders, and foundations in previous chapters we will focus here on the suspension system.

The principle of the suspension bridge is that of a simple clothes line. We stretch a rope between two anchor points and prop the line up in one or more places then hang objects from it. Ropes, chains, or cables are suspended between two towers with vertical (or occasionally inclined) hanger cables attached at intervals stretching down to a stiff road deck. The deck distributes the forces from the traffic on the bridge to the hangers and onto the cables (Figure 32a). The Golden Gate Bridge in San Francisco is a magnificent and iconic example.

Cable-stayed bridges hang in a different way. There are no slung cables between the towers but rather straight cables from the tower to the deck as at the spectacular Millau Viaduct in France (Figure 33). There are two main forms: the fan (Figure 32b) and the harp (Figure 32c), but there are others, such as cable net, and there are many variations on these basic themes. The fan system has cables radiating out from a single point on the tower. The harp has parallel stays from tower to deck. The cable net system has cables connecting the stays to form a network. Some stays are in a single line along the centreline of a bridge. More

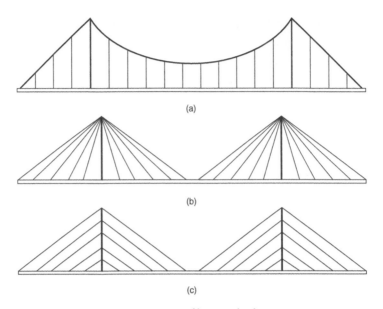

FIG 32. Types of hanging bridges

commonly, two sets of stays to the two sides of the bridge deck
help to resist twisting.

Early builders of hanging bridges with stiff decks experimented with different ways of making sure their bridges didn't wobble. One way was to combine suspension cables with stays—one
of the best known examples is the Brooklyn Bridge in New York
(Figure 37).

Suspension cables may be spun on site. They generally contain
thousands of wires but again there are very many variations.
Modern stay cables and hangers are often made of groups of
strands. The strands are usually spirally wound bundles of wires
of 5 to 7 mm diameter. Typically a wire is four times as strong as

154

FIG 33. Millau Viaduct, France

mild steel but is much less ductile—the strain at breaking may be one-fifth of that for mild steel.

The job of the towers is to create the height from which the bridge deck may be hung—they are the props on the clothes line. They come in many shapes and sizes—most are vertical but some lean. Some are very stiff, usually made of masonry or cast iron, are fixed at their bases and support the main cables through a sliding saddle so the tower only resists vertical compression. Older bridges such as the Brooklyn Bridge in New York and the Clifton Bridge in Bristol are of this type with masonry towers as magnificent portals. Steel portals are at Verrazano Narrows Bridge and the George Washington Bridge in New York. Some towers are hinged at the base and therefore free to rock in the

FIG 34. Pont de Normandie, France

plane of the main cables. They offer no resistance to span direction movements. Many modern towers have fixed bases and attached cables so they are essentially vertical cantilevers. They help to resist movements along the span length.

Some towers are single masts such as the Wye Bridge; some are an inverted Y shape as at the Pont de Normandie (front cover and Figure 34); some have double towers with and without cross beams such as at the Golden Gate Bridge and the Second Severn Crossing or with cross bracing as at the Bay Bridges in San Francisco. At the Millau Viaduct (Figure 33) in France the pylons sit above the deck on separately erected piers below.

There is one important difference we should note between suspension bridges and cable-stayed bridges. In a suspension bridge the main cable tension is concentrated in the anchorages at each end. By contrast the cable tension in a cable-stayed bridge is distributed over many anchorages along the deck. The more stays, the more supports there are to the deck and the lighter and less costly the bridge.

Suspension bridge main cables may be anchored at their ends to earth through anchorage blocks or to piers or to the bridge deck. In some bridges the back stay cables react off piers as at the Bay bridges in San Francisco. Consequently foundations may have to resist large uplift forces. Anchorages are a major issue for suspension bridges, but far less so for cable-stayed bridges.

Stays are usually anchored to towers to make erection easier. Consequently the forces each side can become unequal and so bend the towers as a vertical cantilever. The towers therefore need to be flexible enough to resist.

The horizontal component of the internal tension in the inclined stays of cable-stayed bridges creates a considerable compression in the bridge deck. The decks of many modern bridges are box girders of steel or concrete to give torsional strength and streamlining in wind.

Hanging bridges are not easy to build so the erection process plays a big part in design thinking. Many special techniques have developed—two examples are cable spinning and balanced cantilever construction. John Roebling, designer of the Brooklyn Bridge, New York, was the first to spin the wires into cables for suspension bridges. The modern process is essentially his. Just as the Forth Bridge (see Chapter 2) was constructed by building out from piers in two balanced directions so the same principle is used to construct many modern cable-stayed bridges. The towers, anchorages, and main cables of suspension bridges must be in place before the bridge deck and stiffening girders can be built. Cable-stayed bridges are normally quicker to erect and hence cost less to build.

Hanging bridges are often built in exposed positions so wind loading is often critical both during erection and in use. The wind creates a dynamic force on the whole bridge. Wind and earthquakes can cause problems at all levels, chapters, paragraphs, and sentences.

So how do hanging cables work? In Chapter 1 we looked at direct tension acting axially along the length of a member. With some small differences due to the sagging created by the self-weight of a sloping cable all that we have said will apply to the cable stays of cable-stayed bridges.

However, there are some complications. One of the most important is that cables made up of many wires or strands acting together in tension must be the same length as we shall see later when we see how cable-stayed bridges are erected.

Cables in suspension bridges work quite differently because the loads are applied laterally along its length.

In Chapter 2 we looked at the way a rope or cable hangs between two fixed points as a catenary. Implicitly the weight of the rope was uniformly distributed along its length, i.e. along the curved shape. However, on a hanging bridge with a stiff deck the load from the dead weight of the bridge is uniformly distributed along the span as in Figure 35a—shown by the wiggly line. In this situation the shape of the rope changes from a catenary to a parabola but the distinction in practice is subtle.[3]

Let's imagine what happens to the rope when a large truck moves onto the bridge. You'll recall that bridge builders call this kind of moveable load a live load. In Figure 35b the truck is a

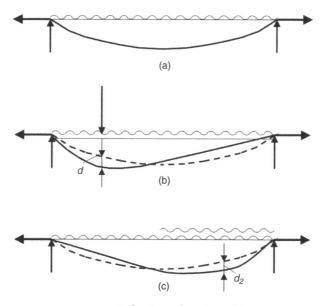

(a)

(b)

d

(c)

d_2

FIG 35. Deflections of tension cable

point load applied directly to the rope as shown. In reality the weight of the truck will be taken by the stiffening bridge deck girder and distributed to the rope through hangers—but we will come to that later. The rope will change shape and the internal tension in the rope will increase. As the ends of the rope are held in place, the horizontal external reaction forces will also increase.

If the sag of the rope gets larger then the value of the horizontal reaction at the ends gets smaller. This is because the ends of the rope will be more vertical and will tend to pull down on the end more. This will increase the required vertical reaction. But what happens if the sag is zero? Theoretically the horizontal force required may become infinitely large. However, in practice this cannot happen. There will always be some sag under a lateral load because the rope cannot take any bending—it has no bending stiffness. A rope can only resist a lateral load by changing its shape.

Figure 35c has a second wiggly line over the right-hand part of the span. Instead of just one arrow as one truck in Figure 35b we now have a train of trucks all in a line—a queue of trucks—hence the new wiggly line. But a line like this can be of any length—how long should we make it? We want to find the answer for the worst case when the deflections of the cable are the largest. You could be forgiven for thinking that this worst case happens when the line of trucks covers the whole span—with a full wiggly line just like the dead load. That is so for a simple beam but a cable is different. The answer is a surprise. It turns out that the deflection of the cable at the centre of the span is largest when the train occupies the middle 40% of the span. The reason for this difference in behaviour between the simple beam and the cable is that there is axial tension in the cable but none in the beam, and the cable has no bending stiffness. Engineers call the stiffness created by the tension in the cable a geometric stiffness.

For most hanging bridges the first wiggly load, the dead load, is much larger than the second wiggly load, the live load. The bridge therefore settles down under the dead load and the changes, as the live loads move across, are smaller.

So far we have simply been talking about the cable. We haven't yet introduced the bridge deck into our thinking and this will make a difference.

So let's add a bridge deck. At the very least this increases the self-weight and hence the dead load of the bridge. In turn this increases the stiffness of the bridge.

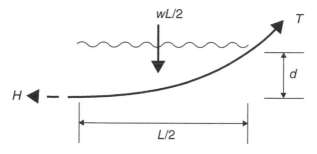

FIG 36. Internal forces in suspended cable

We can begin to see this from Figure 36, which shows the right-hand half of the cable of Figure 35. The cable has been cut in half to reveal the internal force, H—the tension in the cable. At the other end is the external reaction force, T. This is the combination of the horizontal and vertical external reaction forces shown in Figure 35a. Again the wiggly line is the dead weight of the bridge, labelled here w kN/m (i.e. w kilonewtons per metre). Because we are assuming that it is uniformly distributed along the span the total weight on the half-span is w times the length of the half-span, $L/2$, which is $wL/2$, and is shown as the vertical downward arrow. It effectively acts halfway along the half-span, i.e. at the quarter point.

Now let's look at the turning effect of these forces about the right-hand end of the bridge. Recall that the turning effect or moment is a force times a distance from that point or pivot.

The force H acts horizontally and the distance between it and the right-hand end of the bridge is d. Therefore the moment of H about the right-hand end is H times d and it acts clockwise.

The weight of $wL/2$ acts vertically at the quarter point of the span. The distance from where it acts to the right-hand end is $L/4$. Therefore the turning effect or moment of the weight is the total weight of $wL/2$ times the distance $L/4$. This is $wL/2 \times L/4$, i.e. $wL^2/8$ and is anticlockwise.

Now the turning effect of the force H and the weight $(wL/2)$ must balance. This means that $H \times d = wL^2/8$. From this we conclude that

$$H = (w \times L^2)/(8 \times d).$$

This is just a mathematical way of saying three things. First it tells us that the tension in the cable increases directly with the dead load. Second it tells us that the increase is even larger as the span increases—indeed, it increases with the square of the span. So if the span is doubled the tension is quadrupled. Finally the formula tells us that the tension is inversely proportional to the dip—so the larger the dip, the smaller the tension. We noted that earlier by physical reasoning—now we have it through more formal mathematical reasoning.

The ratio of the span to the sag of the cable at centre span L/d is interesting. It is sometimes used as a measure of the form of the bridge and is implicit in our formula for the tension since $H = (w \times L/8) \times (L/d)$.

Brunel's contemporaries knew that if a cable was too shallow the forces became too high but they were worried that if it was too deep the bridge would tend to wobble more. They chose to use ratios of L/d of around 10 for the three iconic bridges of the nineteenth century that illustrate the rapid development in building hanging bridges after Finlay's bridge—Telford's Menai Bridge, Brunel's Clifton Suspension Bridge, and Roebling's Brooklyn Bridge.

Thomas Telford, the son of a shepherd, was born in Glendin-
ning, Scotland, in 1757. He began his working life as a mason and
soon was involved in a masonry arch bridge as well as major
buildings. He left Scotland in 1782 to work in London. In 1787 he
was appointed Surveyor for Public Works in Shropshire. By 1792
his talents for large-scale works were being recognized. He
worked with William Jessop on the Ellesmere Canal in Shropshire and in 1805 they completed the immensely impressive
aqueduct near Llangollen in North Wales, the Pont-y-Cysyllte,
which is still working today. From 1803 onwards he directed the
building of more than 1000 bridges, 1920 km of roads, many
harbours, churches, drainage schemes, water supply, and river
navigation. In 1820 he became the first president of the Institution of Civil Engineers. He died in 1834 and is buried in Westminster Abbey.

One of Telford's landmark projects, opened in 1826, was the
Menai Bridge between the island of Anglesey and the mainland
of Wales near Bangor. Before then the only way to get to
Anglesey was by ferry across what was often a stormy sea.
Largely because of his work in Shropshire Thomas Telford was
chosen by a Board of Parliamentary Commissioners to design the
bridge. The Board was advised by a group of engineers chaired by
John Rennie, who approved Telford's plans. Required to allow a
30-m clearance for sailing ships, he decided on a chain suspension bridge with a 176-m main span and two side spans of 79 m.
The design was worked out in 1817–18 but previous projects, in
particular for a bridge over the Mersey at Runcorn Gap, had led
him to do 200 experiments on 'the tenacity of bar and malleable
iron'. Telford was meticulous about detail. He even built a small-
scale model of the proposed bridge and tested it. He knew exactly

what iron he wanted. The chains were suspended across a nearby gorge in Anglesey in order to test them but also to determine their final lengths and positions. He knew the theory of the time but did not trust it sufficiently not to check it very carefully. He specified that each bar was to be tested to a tensile stress of 170 N/mm^2 and under load would be struck by a hammer.[4] A specially designed machine was built to do this and each test was carefully supervised.

There were 16 chains in four sets—two for each carriageway. The four chains making up each set were made of eye bar links 2.4 m (8 ft) long. The links were held apart by separators and connected by short plates and bolts. The vertical hangers were 25-mm (1-in.), square wrought-iron bars connected to the plates at 1.5 m (5 ft) intervals. Cast-iron saddles on roller bearings on top of the masonry towers carried each set of chains. At their ends the chains were anchored in tunnels driven deeply into bed rock. Every bar was soaked in linseed oil to reduce corrosion.

The timber decking of the first bridge was not stiff and the bridge was damaged by wind even during construction. They connected the two central chains to dampen the movements. A week after opening a severe gale broke some of the cross beams and hangers. About 10 years later there was more damage and in 1839 both carriageways were broken in several places and many hangers lost. However, the chains survived well and the roadway was repaired. In 1893 the wooden deck was replaced with steel. In 1938 the chains were replaced by steel ones and in 1999 the structure was further strengthened.

Telford's difficulties at Menai directly influenced the progress of the building of the Clifton Suspension Bridge, an icon for the city of Bristol. In 1829 a design competition for a bridge over the

Avon Gorge was held and Telford was appointed chairman of the judges. He found fault with all of the entries. His meticulous nature and his difficulties at Menai must have influenced his thinking about the possibility of a longer span. He decided to submit his own design. He proposed building massive towers from the bottom of the gorge. He clearly felt that he had achieved the maximum span possible at Menai. Telford's proposal wasn't popular and so a second competition was held in 1830. The winner this time was I. K. Brunel—only 24 years old, and this was his first major commission.

The foundation stone for the new bridge was laid in 1831 but politics and lack of money caused delays. In 1843 the towers were built but the project was then stopped. Brunel died in 1859 before the bridge was finished. It was eventually opened in 1864.

The total length of the bridge is 414 m with a central span of 214 m. The width between the chains is 6.1 m. The deck is 76 m above high-water level.

The wrought-iron chains are anchored in tunnels in the rocks 17 m below ground level at the side of the gorge. The ends of the tunnels taper out so that the brick infill forms an immovable plug. In 1925 the anchorages were strengthened with fixing stays and chains, and concrete.

In his calculations for the bridge Brunel demonstrated that he knew the latest theory of the shape of the chains including the simple catenary and the shape under uniform loading—the parabola. However, no bridge builder of the time had any appreciation of the stiffening effects of the light longitudinal deck girders. The problem was conceived as the form of the cable—the deck was just a beam slung underneath.

The building of the bridge was started by taking six wire ropes across the gorge and tightening them. Planks were fixed between the wires to make a footway. Then two handrails were installed with another wire at head height. A light frame on wheels travelled across carrying each link of the chain. The wire bridge was the staging for the chains as new links were added. After the first chain was finished, the second was built on top and then the third. Vertical suspension rods link the chains to wrought-iron girders which run the full length of the bridge and divide the footway from the road. Cross girders form the deck structure with Baltic pine timber sleepers 120 mm thick on top and covered with asphalt.

For over a century the abutment on the Leigh Woods side of the gorge was carrying a surprise package. The 26-m-high masonry tower sits on a 33-m-tall abutment. There were no plans and everyone thought that the abutment was solid. Indeed a borehole drilled in 1969 confirmed the idea as it went through solid structure.

However, in 2002, much to everyone's surprise, a honeycomb of 12 vaulted chambers in two tiers, linked by narrow shafts and tunnels was discovered. The largest chambers have a floor area of 17.25 m × 5.6 m and are 10.8 m tall. The upper tier has seven linked chambers. The lower tier stands directly on the rock of the Avon Gorge and has five linked chambers. The chambers were not ventilated but the air was clean with no bats or other creatures— not even spiders. There are stalactites up to 5 m long hanging down.

In 1885 Sarah Ann Henley from Bristol threw herself from the Bridge after an argument with her lover. Not many people survive a fall from that height—76 m. However, she was wearing crinoline petticoats beneath her crinoline dress which, it is said,

FIG 37. Brooklyn Bridge, New York

billowed out, slowed and cushioned her fall. She was injured but recovered and died in 1948 when she was 84.

Unfortunately many people have jumped off bridges. The Brooklyn Bridge in New York—perhaps one of the most instantly recognizable iconic bridges in the world (Figure 37)—is no exception. But the bridge itself was a giant leap of a different kind—technological development. It was declared a National Historic Landmark in 1964 as one of the oldest suspension bridges in the USA. The massive span (for that time) of 486 m connects Manhattan with Brooklyn over the East River. Construction work began in 1870. The bridge was the first use of steel wire for the suspension cables.

John A. Roebling was admirably fitted for the job of designing the bridge since in 1867 he had accomplished the 'impossible' feat of constructing a railway suspension bridge over the Niagara gorge. Earlier that same year he had also completed the Cincinnati–Covington Bridge over the River Ohio with a record span of over 322 m.

A number of people objected to the idea of a bridge and to Roebling's recommendation that steel wire suspension cables should be used. A board of consulting engineers was set up and eventually agreed with Roebling. The proposal for a bridge was approved in 1869 but a clear height of 41 m was specified to allow ships to pass under. The final surveys for the bridge were started in June 1869 but then a major setback occurred. While standing on a fender rack, John Roebling's right foot was crushed by a ferryboat approaching its slip. He developed tetanus and died on July 22. Fortunately his son, Col. Washington A. Roebling, had resigned his army commission in 1865 and was working with his father—indeed, he had taken practically complete charge of the construction of the Cincinnati Bridge. Washington was a natural successor for the Brooklyn Bridge.

Work on the pneumatic caisson for the Brooklyn tower began in 1870 but they soon found that there was a mass of boulders cemented together, making them almost impossible to dislodge. They tried various methods for breaking up the largest boulders but ended up using successively larger explosive charges. One explosion, according to Washington, was 'sufficient to singe off whiskers and create some alarm'.[5] The smoke was 'injurious to the workmen'. Caisson disease—decompression sickness or the 'bends'—was an acute problem because men were taken out of the pressurized atmosphere too quickly. Washington reported that 'scarcely any man escaped without being somewhat affected by intense pain in his limbs or bones or by temporary paralysis of arms or legs'.[6] Although Washington had visited Europe to learn about the disease he himself succumbed when fighting a fire caused by a workman holding a candle near to the roof of a working chamber which was set alight. The chamber had to be flooded to put out the fire but during the fire-fight he had his first attack of the 'bends'. This permanently crippled him and he was confined to his home. Nevertheless he continued to direct the work through his loyal and capable assistants, including his wife. The Mayor of Brooklyn wanted him replaced but Washington refused to budge. After a prolonged and heated debate he was retained by a vote of 10 to 7. (Washington resigned from his post in 1883 shortly after the bridge was opened to traffic.) The New York caisson although founded at nearly twice the depth was sunk in 20% less time since it was founded on sand.

John Roebling said of the towers, 'the great towers will serve as landmarks to the adjoining cities ... The Brooklyn Bridge will forever testify to the energy enterprise and wealth of that community which shall secure its erection.'[7] The masonry towers and anchorages were sufficiently complete by August 1876 to begin installing the galvanized steel wire cables. They used a 'hauling' rope and a 'carrier' rope to install four sets of cables—two at the extreme edges and two only about 4.6 m apart in the middle. The cables were 400 mm diameter with nineteen strands of wires. They were laid on the river bed and then hoisted to the tops of the towers. The remaining ropes were erected by hauling them across the river supported on hangers running on the carrier rope. Access to the work was through a small footway along the length of the bridge with five cradles or crosswalks. The cable wires were spun approximately 18.3 m above their final position at the centre of the span. The cable was spun similar to how it is done now except that only one wire was spun at a time. There was only one major mishap when a wire broke loose and slipped off the New York tower—had anyone been in the way they would have been sliced in half. Unfortunately the strand shoe struck and killed two workmen.

The 24.4-m-wide bridge deck floor structure (Figure 38, top) was originally planned to be of wrought iron trusses 3.7 m deep. However, Washington decided that it would be lighter and cheaper to use six steel longitudinal trusses connected at their base by smaller cross trusses. In order to accommodate extra loading he increased the depth of the four inner trusses to 5.2 m depth and the outer trusses to 2.7 m deep. When the trustees of the bridge expressed concern about the safety of the bridge, Washington replied, 'The margin of safety in this bridge is still over four, which I consider safe.'[8]

In the early 1950s two of the inner trusses were removed and re-erected in place of the two outer trusses to create two 30-ft-wide, three-lane highways between the outer and inner cables (Figure 38, bottom). The flooring was replaced by a 3-in.-deep, concrete-filled steel grid. In 1999 this was replaced by precast steel and concrete panels.

Experience had demonstrated to Roebling that the cable had to be stiffened against oscillatory motions caused by moving loads or by the wind. There were two different methods of achieving this. One was by using inclined stays and the other by stiffening girders along the floor—he used a combination of the two. The trusses stiffened the cables vertically in the middle of the centre span and resisted wind. The stays, varying from 45 to 51 mm in diameter, were used to reduce the motions of the cable saddles under unbalanced loads on the centre and side spans. The stays

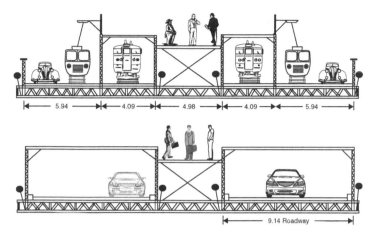

5.94 | 4.09 | 4.98 | 4.09 | 5.94

9.14 Roadway

FIG 38. Cross sections of Brooklyn Bridge

do relieve the cables of load but are part of the unique feature of
the architecture of the Brooklyn Bridge. The side spans are
unusually large—it is not clear why.

The influence of the bridge deck and stiffening girders were
still not well understood even when Roebling built Brooklyn
Bridge. After it was completed, some major steps were made in
the theory of suspension bridges. The important stiffening effect
of the deck girder began to be realized. Telford, Brunel, and
Roebling were all familiar with the latest theoretical developments but they were pushing the boundaries. Figure 39 demonstrates the complexity of the problem. Again the wiggly line
along the deck of the bridge represents a uniformly distributed
load—shown across the whole length of the bridge. The live load
will normally cover only part of the span, as in Figure 35c.

In Figure 39a a suspension bridge has been cut through vertically to expose the internal forces. Compare this diagram with

Figure 26 for the statically determinate Pratt truss. A suspension bridge is statically indeterminate because there are more than three unknown internal forces on each of the two separated parts. We know the load but we have six unknown forces. Two of them are external reactions—the reaction at the base of the tower and the anchor force needed to hold the cable down. Four of the unknowns are internal forces in the cable and the deck.

In Figure 39b the bridge has been cut through horizontally to expose all of the many internal forces in the hangers (tension) and the towers (compression). They can't easily be calculated unless we make some gross simplifications.

The first attempt at a theoretical analysis was carried out in 1858 by Glasgow University Professor William Rankine. Born in Edinburgh, he was one of the founders of the science of thermodynamics and worked on a whole range of topics from steam engines to bridges, prolifically publishing hundreds of technical papers. His engineering manuals were used for many decades. In order to make progress on the analysis of suspension bridges he made three simplifying assumptions. First he said that under dead loading the cable is parabolic and the bridge deck girder is unstressed—in other words, the upward pulls of the hangers effectively neutralize the downward dead weight of the bridge. Second, any live load on the deck is distributed uniformly by the hangers to the cable. Third, the effect of the hangers on the deck girder is to apply an upwards uniformly distributed load which is the live load divided by the number of hangers.

Rankine made the problem statically determinate so that all of the internal forces could be determined by balancing the known ones with the unknown ones. He found the force in the suspension cable using the calculation described earlier with Figure 36.

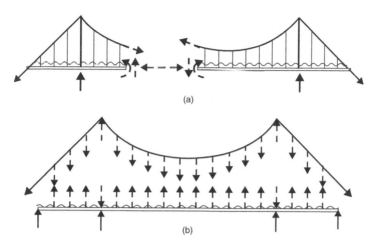

(a)

(b)

FIG 39. Internal forces in a suspension bridge

He balanced the vertical forces, then the horizontal forces, and then the turning effect of the forces, i.e. taking moments.

Later researchers replaced Rankine's third assumption by the idea that the upward uniform distribution of loads from the hangers depends on the stiffness of the cable in tension and the stiffness of the deck girder in bending. The structure was still assumed to be linear elastic (see Chapter 2) but the calculation became statically indeterminate and hence a lot more complicated. The principle used was that the increase in cable tension due to the live load was found such that the total strain energy in the bridge is a minimum.

You will recall that we referred to strain energy in Chapter 3. In fact the idea was first suggested by Daniel Bernoulli in 1742. He said that the shape that a bent beam takes up is the one for which the strain energy is a minimum. The idea turned out to be of

173

immense importance. It was used almost immediately by Leonhard Euler in 1744. Leonhard was a pupil of Daniel Bernoulli's father, John, and one of the most eminent mathematicians of the period. Daniel challenged him to find the buckling load of an elastic column and the result was the buckling load that we came across in Chapter 4.

Much later, in 1873, Alberto Castigliano produced his strain energy theorems (see Chapter 3) as part of his doctorate at the Polytechnic of Turin. But not without controversy. Luigi Menabrea and Alberto Castigliano disputed who had the ideas first. Castigliano's results contain the principle of least work (see Chapter 3) as a special case. Menabrea was the older and much more senior man. He had been Prime Minister of Italy (1867–9) as well as a distinguished academic. He had published least work principles for elastic bodies and his work contained the principles that Castigliano developed and verified. But in 1875 when Menabrea published a new paper and referred only to Castigliano's work in a footnote, Alberto was incensed and objected strongly. The Accademia Nazionale dei Lincei based in Rome was asked to intervene. Founded in 1603, the Accademia was one of the foremost scientific authorities in Italy—Galileo was one of its first members. The committee found that Castigliano had 'done a good piece of work' but that Menabrea had made the principle popular and useable. Regardless of the committee's conclusions it is Castigliano's name that has been handed down to generations of students.

In 1888 an Austrian, Josef Melan, put forward a more advanced 'deflection theory' for suspension bridges. He argued that the deflection of the girder was significant. The mathematics of differential calculus developed by Newton and Leibnitz in the mid-seventeenth century was used to produce so-called 'differential equations' to model the forces flowing through a suspension bridge.[9] Unfortunately the theory rapidly became too complicated for anyone but specialists to follow. Various approximate solutions were developed, and important mathematical solutions and techniques produced, particularly by Stephen Timoshenko, who was born in the Ukraine in 1888 but worked as an academic mainly at Stanford University in the USA from 1936 to 1964. However, it wasn't until English mathematician Richard Southwell at the University of Oxford in 1939 began to apply his new numerical methods[10] to the problem that the modern revolution in calculation techniques took off. His methods relied on hand calculations but they were the forerunner of the powerful computer techniques of today—as we will see later in the chapter.

These developments in the theory of hanging bridges are just one example of the growing influence of mathematical and scientific reasoning in the physical sciences and technology over the past two hundred years. The influence has become so strong that many refer to technology as applied science. Technical success has reinforced the view that objects are 'things' quite distinct and separate from humans.

We, of course, recognize that we are made of atoms and molecules. But we know that our conscious minds and our sense of being and mystery make us different. To many people the distinction is a simple statement of the obvious. As a result, ideas held before the relatively recent successes of science seem primitive and not worthy of consideration. Of course we would not want to sacrifice virgins to the river gods anymore (see Chapter 1) but the idea that we are not part of the physical world is so misleading as to be dangerous (see Chapter 4).

Another example of this kind of thinking, which may seem at first to be rather trivial, turns out to be of profound importance for bridge builders. We routinely define a product as an object, and a process as what is done to make that object. However, an everyday product like a kettle is an object that does not have a fixed and immutable state—it is conceived, designed, manufactured, used, and destroyed—a process of change through time. The same is clearly true of bridges.

The distinction between processes and products is useful but can be distinctly unhelpful if it causes a neglect of change. In the past, most bridge builders were commissioned only to design and build a bridge. The bridge was delivered, and nothing more. Not enough thought was given to long-term effects. There are even instances where designers weren't involved in the erection of their bridges. Maintenance and decommissioning wasn't even on their radar. The problem has been exacerbated by the highly fragmented nature of the modern construction industry brought on by the enormous increase in specialization. The overview that Telford, Brunel, and the Roeblings were able to maintain was, and still is, often lost. With no 'guiding mind', projects can lack 'joined-up' decision-making. These ideas will be explored further in Chapter 7, since they have implications far beyond building bridges.

As we have already noted, Telford, Brunel, Roebling, and the other great bridge builders of the nineteenth century were pushing at the very boundaries of what was possible. They took cognizance of the science of the time but realized its limitations. The science of bridge building was still relatively underdeveloped and its claims were modest. But as science became more obviously important, bridge builders had to take the time to learn and understand. People had to specialize—there was just too much for any one person to fully grasp. The challenge for the bridge builders of the twentieth century was to integrate the new scientific developments with practical achievement. Inevitably technology submitted to the intellectual culture of universities with profound consequences, which will be explored further in Chapter 6 when we look at the role of science in practical decision-making and at its impact on our ideas of risk.

At a more detailed level bridge builders had to simplify their problems to use the science—just as Rankine did for his first attempts at calculating the internal forces in suspension bridges. Historically bridge builders have tended to think of bridges as static objects. Hanging bridges force a recognition of time because they are so clearly dynamic changing processes.

It may seem obvious that internal forces will change as traffic crosses a bridge. But going back to our masonry arches of Chapter 2, the dead load weight of the bridge itself so dominates the thrust line that any changes due to the live load are hardly noticeable. That is why masonry arch bridges are so robust—change is very small and very slow.

The beams and trusses of Chapters 3 and 4 were more complicated, and robustness had to be consciously sought. Larger spans could be achieved but the bridge builder had to ensure that their designs made the bridges safe. Many bridges are statically indeterminate; thus, to find the flow of internal forces, bridge builders had to use scientifically advanced theorems of strain energy. But the theories developed focused on statics—the state of a structure when forces do not vary over time and all action forces are resisted by sets of equal and opposite reaction forces.

There was little explicit recognition that change through time is important.

So how did this go unrecognized for so long? If time is so important why weren't there more failures? A theorem first suggested in 1936 by a Russian, Alexei A. Gvozdev, and later developed by William Prager, called the 'Safe Theorem' tells us why. Apart from its advocation by Jacques Heyman, it has not received the wide attention it deserves.[11] You will recall in Chapter 2 that we didn't need to know the exact position of the thrust line in an arch bridge; just that it was in the voussoirs. The Safe Theorem states something similar: as long as we can find a set of forces in the bridge in equilibrium and which nowhere exceeds the strength of the bridge, then we can be confident that the bridge is safe. This is why we can design bridges without always knowing the exact internal forces at all times. In effect the Safe Theorem states that as long as we deal with the extremes of maximum loads and minimum material strengths, we can forget the processes of change in between.

But hanging bridges remind us that this isn't always so when time is important. There are four important ways in which time is a big factor in how a bridge behaves. They are vibration (including flutter), impact, fatigue (including fretting, see below), and deterioration (including corrosion). The timescales, however, are very different—varying from seconds (flutter) to years (corrosion).

All bridges vibrate, but hanging bridges are particularly sensitive, as seen with the wobbly London Millennium Bridge at the very beginning of this book. The main causes are traffic (pedestrians, cars, trucks, and trains), wind, and, in some parts of the world, earthquakes. Vibrations, like all of the movements we have looked at so far, occur in the various degrees of freedom.

They can happen at all levels of a bridge. So a whole bridge can vibrate, or just individual parts, such as the suspension system, a particular truss in a bridge deck, or a cable hanger.

Every bridge and every part of it has a particular natural frequency of vibration, just like a plucked guitar string. The string vibrates at its natural frequency, moving freely in a degree of freedom—lateral displacement. A thin diameter string with low mass vibrates quickly, i.e. a high frequency giving a high pitch note. You tune your guitar by adjusting the tension, which changes the natural frequency. The other way to change the note is to change the length of the string, which is what happens when you press your fingers on the finger board to play a tune. The lower notes come from larger diameter, heavier strings, which vibrate more slowly so you get a low pitch note or a low frequency.[12] The natural frequency of a bridge depends on similar factors—its mass (equivalent to the diameter and weight of the guitar string) and its stiffness (equivalent to the tension and length of the guitar string).

If a vibrating force (like the wind) is acting on a piece of the bridge and both are moving at the natural frequency of that piece of bridge then the vibrations can become very large. As we saw in Chapter 1 this is called resonance. It is similar to the effect you get when you push a child's swing. Very quickly you learn when to push for maximum effect—in effect you have tuned yourself to push at the natural frequency of the swing.

So how do bridge builders avoid these vibrations? They must design and build every part of the bridge to avoid resonance. Unfortunately they don't have much control over the natural frequencies since they depend on mass and stiffness and they can't change the natural forces from the wind or earthquakes.

However, they can influence the effects of these forces—for example, making the shape of the bridge deck more aerodynamic or streamlined influences the interaction between the wind and the bridge.

Fortunately there is another kind of help—damping—as we saw for the London Millennium Bridge in Chapter 1, but it isn't straightforward. Damping is the degree to which the bridge has built-in shock absorbers, which absorb energy and reduce the levels of vibration naturally. If you stop pushing the swing, the amount of to and fro movement (the amplitude) gradually reduces and the swing eventually comes to a halt, mainly through friction in the hinges and the resistance of the air. The same effect occurs in bridges. Friction between different parts of a bridge as they rub against each other will dampen vibrations. There is also an internal friction within the materials and, of course, air resistance. Bridge builders must assess what degree of damping there might be in order to understand what might happen to their bridges and to limit resonance. But it isn't easy, as the experience of the wobbly bridge shows—although the problem there was the unexpected pedestrian loading, not just the low damping. However, the solution, as we saw in Chapter 1, was to increase the damping. The damping force is usually assumed to be proportional to the velocity of the moving parts.

Perhaps the most famous of all bridge vibration stories is the collapse of the Tacoma Narrows Suspension Bridge in 1940. This was an event that revolutionized the way structural engineers thought about the effect of wind on large slender bridges. The bridge, nicknamed 'Galloping Gertie', was filmed as it collapsed—a movie that has been viewed by millions as it has been screened regularly on TV over the years. As luck would have it, the camera wasn't operating at a crucial stage so part of the collapse was missed.

The bridge spanned 853 m. The stiffening deck consisted of two solid plate girders 11.9 m apart. The investigators into the collapse found that the bridge was well designed and built to resist safely all static forces. The designer, Leon Moisseiff, was a leader in his profession and the quality of materials and workmanship was high.

Despite this, the deck started to oscillate up and down as the bridge was being erected. Similar oscillations had been seen previously, including at the Menai Bridge in 1826. The stresses in the Tacoma Narrows Bridge must have been high at times but there was no evidence of any consequent structural damage. However, after four months, just before it collapsed, the nature of the oscillations suddenly changed. A cable connection slipped at mid-span, the movements increased, and the deck began to twist as it vibrated. They continued to get worse and the vertical hangars began to break. Progressively the entire structure collapsed. All of this occurred under a steady wind speed of only about 42 mph.

If the bridge was designed properly, why did it fail with such a low wind speed? Many textbooks refer to the bridge as a classic example of forced resonance where quite simply the wind gust frequency matched the natural frequency of the bridge.

Yet wind engineers now know that this explanation is inadequate, since the wind speed was steady. So what was the explanation? Table 1 offers some clues. It lists some of the large suspension bridges being built in North America around that time. Two ratios for each bridge are listed—the span to width of the bridge and the span to the depth of the stiffening girder. The ratio of span to width is a very crude and approximate measure of the torsional characteristics of the bridge.[13] Note that the figure for the Tacoma Narrows Bridge was much higher than any of the others. The span to depth of the stiffening girder is a crude and approximate indication of the longitudinal bending stiffness of the bridge. Again the Tacoma Narrows Bridge had a much higher figure than any of the others.

However, it was the decision by Moisseiff to use solid plate girders for the deck that proved to be the main problem. By that decision he created an H-shaped girder with bluff (nonstreamlined) surfaces in the wind. We now know that this is one of the worst aerodynamic shapes that he could have chosen. As the wind flows around a bluff body the air breaks away from the surface and moves in a circular motion like a whirlpool or

Table 5.1 Some suspension bridges in the USA around the time of the failure of the Tacoma Narrows Bridge.

Name	Year	Span (m)	Span/width	Span/depth
George Washington, New York	1931	1067	33	120
Golden Gate	1937	1280	47	168
Lions Gate Vancouver	1938	472	39	104
Bronx-Whitestone New York	1939	701	31	209
Tacoma Narrows	1940	853	72	350

whirlwind as eddies or vortices. Under certain conditions these vortices may break away on alternate sides, and as they are shed from the body they create pressure differences that cause the body to oscillate. The low torsional stiffness and hence the low natural frequencies just made this bad situation worse.

But this is not the whole story. A pertinent question is—did the shedding of the vortices cause the motion or did the motion cause the vortices? There is still controversy over the details of what caused the final catastrophic vibrations. However, most wind engineers now say it was 'self-excited flutter'. It was a behaviour that emerged from a complex interaction between the wind and the bridge deck. Let me try to explain.

We know that there must be a balance between the external forces on the bridge and the internal forces within it. The external disturbing forces on the bridge came from the wind. In overview, as the bridge moved, there were three internal forces—the inertia due to the mass of the bridge, the restoring force due to the stiffness of the bridge, and the force due to damping.

Flutter occurred at Tacoma when the disturbing force of the constant wind speed on the bridge was controlled by the motion of the bridge. If the motion had stopped, the disturbing force would have stopped too.

An arbitrarily small rotation of the bridge deck changed the angle of the wind to the bridge (called the angle of attack). The twisting increased the angle of attack of the wind, which caused more twisting. The two ends of the bridge twisted in opposite directions defining what engineers call the 'mode shape'—the shape of the twist along the span.

As the twisting angle grew larger, the restoring force due to the torsional stiffness of the bridge increased too. Eventually this restoring force overcame the disturbing force and the bridge began to rotate back again.

As it rotated back the restoring force reduced. However, the inertia force and low damping meant that when the bridge got back to its starting position it carried on in a reverse twist. As it did so the disturbing force of the wind again started to increase the twist but this time in the opposite direction.

Eventually the restoring force again overcame the disturbing force. The reverse twisting slowed and stopped and the bridge began to twist back again.

This cycle of events, twisting one way and then the other, continued for 45 minutes. But with each twist there was effectively positive feedback because with each cycle the angle of twist increased. The bridge was effectively in a state of self-excitation. Eventually the twisting became so severe that the internal forces overcame the internal strength of some of the bridge components and the bridge shook itself to pieces in a very dramatic way.

Remember that this all happened with a constant wind speed. The wind force did not vary with time—only with the motion.

In a lecture to the Institution of Civil Engineers in 1977 Paul Sibley and Alistair Walker showed how, since the end of the nineteenth century, the theory developed by Melan had enabled bridge builders to design much lighter and more slender bridge decks. Early suspension bridges had been proportioned intuitively and empirically. Moissieff seemed unaware of the possibility of large vibrations and chose plate girders instead of the usual truss to economize on materials.

The Tacoma Narrows collapse is a classic example of failure due to a mode of behaviour not really understood by the technology of the period suddenly becoming important. There was a step change in the two basic parameters shown in Table 5.1—the ratios of span to width and of span to depth. This, combined with the use of a different form of deck construction, was enough.

This type of collapse is very difficult to predict. There were warning signs but they weren't recognized. The difficulty of the problem is perhaps best appreciated by asking 'What warnings signs of future problems are there now?' We will examine this question again in Chapter 6 when we look at risk.

Vibration is the first way in which a bridge is clearly a time-dependent process. The second way is through an impact or the sudden application of load. No matter whether a bridge is beam, arch, truss, or hanging, when a truck crashes into a pier, the truck usually comes off the worst—but both may be damaged. Drivers of double-decker buses have been known to misjudge the safe height of a low bridge with inevitable consequences. A big truck can remove completely an important part of a bridge such as the pier in a central reservation of a motorway or freeway supporting an over bridge. As vehicles cross over a bridge there are small impact forces as they ride over rough surfaces.

Impact is a very short sudden force—an impulse. Because of the velocity at impact the force can be many times more than the dead weight of the truck. The simplest way to estimate the force is to use what is called an impact factor. This is the number of times the static load must be multiplied by to get the impact force on the bridge and it may be as large as 10. Nowadays we use finite element methods (see later in this chapter) to study impact but it is not easy as the response is short-lived or transient.

Perhaps the most common and hence most worrisome time-dependent phenomenon is fatigue. Again this applies to all types of bridge. The word is descriptive—under certain conditions the materials just get tired and crack. A normally ductile material like steel becomes brittle. Concrete is already brittle but cracks can form and grow. Fatigue occurs under very small loads repeated many millions of times.

All materials in all types of bridges have a fatigue limit. If, each time a car or truck travels over the bridge, the stress is below that limit then fatigue damage will not occur. Above that limit, even though the stresses are still very small under certain conditions, each car and truck will cause unseen cracks to form and grow a little.

Fatigue damage occurs deep in the material as microscopic bonds are broken. The problem is particularly acute in the heat-affected zones of welded structures. This is because welds tend to have microcracks that can't be seen with the naked eye. The welding process (see Chapter 4) causes metallurgical changes that can make the steel more brittle, though the effects can be minimized by a good welder. Very strong high carbon steels tend to be less ductile than ordinary mild steel—they are therefore more vulnerable to fatigue cracking.

Engineers look for a relationship between a stress level in a particular material like steel and the number of cycles to cause fatigue cracking. They test specimens of material or specific parts of a bridge by loading them many millions of times—usually in a laboratory rig. They vary the stress levels and plot a graph of the number of cycles to failure for a given stress level. The graph is known as an S-N diagram, where S stands for the stress level and N the number of cycles to failure. Unfortunately there is a lot of scatter in the graph, making them difficult to use—nevertheless, they are often used to calculate an estimate of how long a bridge can last before there is significant fatigue cracking—its fatigue life.

A related phenomenon is fretting. This happens when two surfaces in intimate contact wear as they rub over each other. It can also increase corrosion. Unfortunately, it can happen in the cables of major suspension bridges such as the Forth Road Bridge, and the wires break. This may eventually lead to enforced closure.

Fatigue and fretting are cumulative damage as cracks grow deep inside a material. It is a kind of deterioration—a kind of decay. Decay of a more general kind becomes a big issue when bridges are neglected and maintenance is not as it should be. The story of the Point Pleasant Bridge illustrates what can happen.

In December 1967 the US-35 highway bridge over the River Ohio, linking Point Pleasant, West Virginia, with Kanduga, Ohio, suddenly collapsed. The whole process took only about one minute. Thirty-seven vehicles were on the bridge at the time and 31 fell with it, killing 46 people. The 213-m-span bridge was built in 1928. It was an unusual bridge in that it was suspended from steel 'eye' bar chains acting as the top chord of the stiffening trusses for about half their length. The term 'eye' just refers to the holes at each end of the bar.

The eye bars were between 13.7 and 16.7 m long, 305 mm wide with varying thicknesses of around 50 mm. They were erected in pairs so that at any joint there were four eye bar heads connected by a pin. The steel had quite a high carbon content so that it was stronger but potentially more brittle. The eye bars were designed to fail in the shank rather than the heads. The bridge collapse was triggered by a fracture at the head of an eye bar which had reduced in cross section because of a crack that had grown over a long time. After one eye bar failed, the pin rotated and the other eye bars fell away. It was found later that the crack was due to stress corrosion and corrosion fatigue.

The formal inquiry into the collapse found no mistakes or significant errors in the design and construction of the bridge. There was a minor error in the computed dead load stress of one member which was of no significance. The stresses in the structure at the time of collapse were well below those permissible in the design.

When the bridge was designed, stress corrosion and corrosion fatigue were not known to occur in the class of steel used under the conditions of exposure found in rural areas. The steel was in accordance with the specification but we now know that fractures can be propagated at low energy levels compared to those required in the ductile range. They had grown at a section which was not accessible for inspection and next to a pocket where water collected. The joints would have to have been taken apart to find the cracks.

Several trends converged to cause the failure. First, higher strength steels with higher carbon content were being used. Second, higher permissible stresses were being used more and more as confidence in the applied loading was high. Third, bridge builders were not computing secondary bending effects or indeed any local effects. Fourth, small cracks through stress corrosion had been known in only a few metals under severe exposure situations. Finally, because of the way the links were connected, if one failed, total collapse was inevitable.

Maintenance and inspection of this bridge was a crucial factor. The maintenance authority had a maintenance manual available for use from 1941 onwards for the bridge; however, there was evidence that this was not used during bridge inspections. A complete examination of the bridge was made in 1951. Other inspections were made periodically with emphasis on repairs to the bridge deck, sidewalk, and the concrete of the piers.

Again this example shows the difficulty of predicting the likelihood of some failures. The bridge was built without any major error, to specification, and operated successfully for 39 years before sudden collapse. Although the maintenance left something to be desired, the crack which initiated the failure could not have been detected without dismantling the relevant joint. The detailed maintenance and monitoring of such a structure is obviously essential.

The story of Point Pleasant Bridge illustrates the responsibilities of the people who maintain some of the most important bridges in the world.

Responsibilities don't come much bigger than looking after the Golden Gate Bridge in San Francisco. On my second visit there in 1989 my wife and I experienced one of the most frightening ways in which bridges can be loaded—an earthquake. We had just arrived and had gone to bed when we were woken by the TV in the hotel room shaking violently. The next day we learned that one person had died. Later that year in October an even bigger quake (6.9 on the Richter Scale) killed 62 people, destroyed many buildings, and the two-tier Bay Bridge and Nimitz freeway both partially collapsed. Television footage was dramatic with pictures of cars perched on the edge of a deck that had fallen from its supports. The Golden Gate Bridge was unharmed, but imagine being responsible for checking for possible damage.

Earthquakes shake the ground because there is a sudden movement in the Earth's crust. Tens of thousands of earthquakes occur every year. Some are tiny tremors and some cause major disasters. The ground has the same six degrees of freedom as bridges: movements side to side, back to front, up and down, and rotating in three planes (Chapter 1). Again the need is to limit resonance as ground vibrates in its complex way.

The Golden Gate Bridge was completed in 1937 after four years of construction. It carries US Highway 101 north from the peninsula of San Francisco to Marin County, the wine country, and north California.

The bridge is instantly recognizable from photographs because it has always been painted in an orange vermilion colour and is often shrouded in mist, giving it a somewhat mystical appearance. It is over 1.96 km long. The main span between the towers is 1.28 km. It is 27 m wide and the deck is 67 m above the high water level. It can move by 3.3 m at centre span and swing sideways by 8.4 m. The main towers rise 227 m above the water level. They deflect sideways by 0.3 m and along the length of the bridge by around half a metre.

Eleven submissions were made to build the bridge. The proposal from Joseph Strauss was selected in 1929 and he was named as Chief Engineer. Leon Moisseiff, O. H. Amman, and Charles Derleth, Jr., were Consulting Engineers. Strauss took safety seriously. He insisted on protective headgear, and special hand and face creams to protect against the wind. He had a safety net suspended under the bridge—the nineteen men who fell into it became members of the 'Half-Way-to-Hell Club'. There was only one fatality up to 1937 but then unfortunately ten men were killed as some scaffolding fell through the safety net.

The main cable is nearly a metre in diameter with over 27,000 galvanized wires of 4.9 mm (0.192 in.) in diameter. The wire in each cable was laid by spinning the wire using a loom-type shuttle that moved back and forth as it laid the wire in place. It took 6 months and 9 days. The south tower foundation is 34 m deep. The two main cables which pass over the tops of the two main towers on saddles are secured at either end in massive anchorages.

The grammar of the book of the Golden Gate Bridge was deflection theory with the developments mentioned earlier by Timoshenko. Since then our grammatical ability to analyse large bridges has been revolutionized by the use of computers.

Throughout the book we have explored the internal forces within a bridge by cutting out elements and balancing out the forces. Statically determinate structures with three degrees of freedom can be calculated reasonably easily because all we need to do is balance the external and internal force vertically, horizontally, and turning—up and down, side to side, and rotating. We also know that if a structure is statically indeterminate then we have some energy and virtual work theorems to use. However, it turns out that the mathematics of the equations is often not easily solved—especially for practical bridge structures. To solve these we have to resort to approximate methods which, using modern computers, turn out to work really well.

You will recall that the relationship between the internal forces and stresses on an element and the consequent displacements and strains is known as the stiffness. Force equals stiffness times displacement. So far we have just thought about each one of these separately; for example, the rope of Chapter 1 in tension stretches and the ratio of the stress to the strain is the stiffness of an element of the cross section of the rope. However, a direct tension can also produce a lateral strain because as an element stretches its volume hardly changes so it gets a little bit thinner. If there are three degrees of freedom then every force could have an effect in one of the other degrees of freedom so there are nine possible relations between the forces and movements. Mathematicians write and manipulate these relationships using a special form of mathematics called matrix algebra. We won't go into any details except to note that this makes it possible to write down equations at different levels of the bridge's structure. So just as we can talk about the stiffness of a rope in direct tension so we can talk about the stiffness matrix of an element with all its three or six degrees of freedom.

Based on this kind of idea a whole new subject called numerical analysis was born. There are many techniques used but one of the most important is the finite element method. In it the bridge is modelled as a set of interconnected elements. For a truss the elements might be the individual struts and ties. For a concrete slab the elements might be virtual triangles or rectangles. In effect the slab is cut up into triangles or rectangles as we did for the rope in Chapter 1, the arch in Chapter 2, the beams in Chapter 3, and the truss in Chapter 4. However, the internal forces between the elements are now many more and it's not obvious how they work.

In order to find them some approximations are made. These have been found to work extremely well and have allowed bridge builders to successfully solve some very tricky problems (especially important in cable-stayed bridges as we will see in a moment).

The analysis proceeds in three steps. Firstly, the slab is idealized, as I have described, into these finite elements. However, the first approximation is that the elements are assumed to be joined only at the corners or nodes. So each triangle is joined to neighbouring triangles at its three nodes. Secondly, a mathematical relationship is assumed between the movements of the nodes and the internal movements within the element. Using this assumption, a stiffness matrix which expresses the relationship between the forces on the element with the nodal movements is derived. The third step assembles all of these finite elements together and uses the computer to solve the resulting equations and then calculates the movements of the elements. Once this is done the computer calculates the internal forces. Modern bridge builders can use purpose-made computer software packages to do these calculations for them. But they still need to understand what these programs do, the theoretical assumptions that have been made, and the context in which they can be used. There is still a considerable need for judgement.

Many modern big hanging bridges are not suspension bridges but instead are cable-stayed. As I said earlier they work quite differently from suspension bridges. The internal forces are depicted in Figure 40 in a manner similar to those of the king-post truss in Figure 24, where the wiggly line is again the combined dead and live loads shown as acting across the whole span. Cuts have been made in three places to expose the internal forces. The forces shown are indicative since the actual values will depend on the detailed geometry and loading. At the left-hand support the vertical and horizontal external reactions holding down the bridge are the solid arrows. These forces must be balanced by the upward forces of the cables in tension, the applied dead and live load (the wiggly line), and the internal forces in the bridge deck as it bends as a beam, i.e. bending moment, shear force, and axial compression. The bridge builder must calculate their values under the many possible loads that may come onto the bridge as well as the many hazards such as wind and earthquakes that may threaten it (for more on this see Chapter 6).

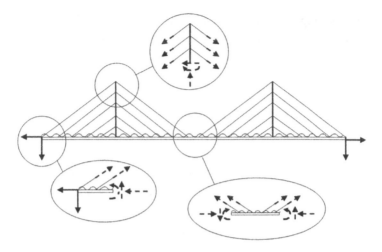

FIG 40. Internal forces in a cable-stayed bridge

The section cut from the centre of the bridge again shows the upward tensile forces from the cables, the downward forces from the dead and live load, and the internal forces in the beam bridge deck. The section cut at the top of the mast shows the internal tensile forces in the cables pulling down and the upward internal compression force in the pylon or mast together with bending and shear.

Modern bridge builders know that they can analyse the bridge reasonably accurately using finite elements. However, one of the difficult problems for cable stayed bridges is that their erection is not straightforward.

If you have ever erected a tent with rope stays you will know the problem. Just as you tighten one stay another loosens. The tensions in the stays are interdependent. If you change one then you change another. However, the situation is worse than that. If you get the tensions wrong your tent (or bridge) could collapse.

Imagine you have two strings of similar but slightly different length of around half a metre. You grab the ends and pull. Two strings are twice as strong as one string—correct?

Unfortunately not. The situation is a bit more complicated than that. As you pull the two strings the shorter string goes taut but the longer one remains slack. All of the force you are putting into the two strings is going into only one string. If you pull so hard that the string breaks then your pull gets transferred suddenly to the longer string—but that will break too. You only get the strength of both strings if they are of exactly the same length and hence take half of the pull each. In a cable-stayed bridge the problem is much harder since the cables are all of different lengths but all must take their share of the forces holding up the bridge deck.

So when bridge builders erect the deck girder of a cable-stayed bridge they must be very careful to get the tensions in the stays right. One way they do that is to use the finite element method on a computer to simulate the erection sequence in reverse.

The first step is to decide what the final profile of the bridge should be when all of the cables are doing their share of the work. The bridge designer then analyses the whole completed bridge with that profile with only dead load acting. He uses the computer to find all of the internal forces in every part of the bridge including the tensions in each cable.

The next step is to simulate the bridge just before the final pieces of bridge deck are put in place. He does this by removing some of the bridge deck girder at centre span in his computer model and reanalysing the bridge. The tensions in the cables will have changed. He notes the tension in cables in the centre because these are the values that those cables need to be tensioned to just before the final bit of the bridge deck is erected (i.e. the bit he has just removed) during the actual erection sequence.

He then removes the two most central cables and another section of the bridge deck. Again he reanalyses the bridge and finds the changed tensions in the innermost cables. He notes that these are the tensions that that cable must be stretched to at this stage in the actual erection sequence, i.e. just before this section of bridge deck (the one he has removed in his analysis) is actually erected.

He then removes the next set of cables and more of the bridge deck and reanalyses the bridge. Again he notes the tensions in the cables because these forces will have to be jacked into them in the actual erection sequence.

He repeats this process until he gets to the cables nearest the towers. These are the first to be installed in the actual erection.

So in the actual erection process each cable is tensioned to the levels found in the computer reverse analysis. If the model is a good one then when the bridge is finished the final profile will be the one desired. The closeness with which the actual final bridge profile and the internal forces match their intended values depends on the accuracy of the reverse analysis and the various tolerances in the construction process.

This is all under dead load. Live loads will cause different internal forces which must be such as not to threaten the safety of the structure. Clearly the bridge builders can check all of the load combinations to check the safety of the bridge (see Chapter 6).

The erection sequence is both simple and complex. The basic ideas of the ways in which hanging bridges work are simple—the clothes line of the suspension bridge and the straight stays of the cable-stayed bridge are reasonably straightforward. However, it is the interdependent connectivity of the cables that makes the task of hanging bridges complex. That complexity makes it difficult to calculate the internal forces, control vibrations, and design erection schemes.

Cables make complexity in other ways too. One is that cables create artistic opportunities that are hard to resist. Art is a different level or type of complexity. Art is 'soft'—by that I mean 'not clearly definable'. Art is about emotional reaction. It is the very opposite to 'hard', by which I mean 'clearly definable and measurable'. The engineering science of hanging bridges is hard, the aesthetics of bridges, whatever their form, is soft—but more on these distinctions in Chapter 7.

Cables enable us to build bridges with a lightness of touch that can be beautiful. Indeed they enable us to make bridges that are not simply beautiful but can lay claim to be conscious public works of art.

The light and shallow curves of the cables of large bridges like Golden Gate, Humber, or Tsing Ma demonstrate the principle of the beauty of harmony between form and function. But, I will argue, they are not works of public art. Their undeniable beauty derives naturally from their structural form in a naturally harmonious context. It is almost an unintended consequence of their structure—the flow of their internal forces. Their aesthetic quality is like that of many garden plants—a naturally evolved beauty through form. If you hold the view that artistic beauty can derive from natural form, then they are beautiful—but are they works of conscious public art? Their beauty arises from their principal duty—to be an efficient structure.

So what is the difference between a bridge which qualifies as public art and one that doesn't? A firm definition of art is elusive. Nevertheless art can be created and enjoyed without one. People make art but also decide what objects do and do not qualify as art. One thing is clear—our decisions depend on time. What experts consider as high art changes through time—there are few absolutes although some argue that our aesthetic taste may have been moulded by evolution. As we said in Chapter 1, since Kant, taste has been considered to be the ability to see harmony detached from personal interest—but this seems inconsistent with the diversity of different cultures.

Classic definitions often refer to two basic elements—knowledge and production. Art was knowledge of the rules for making things—but also the capacity for making something—a power of the practical intellect. So designing combinations of naturally beautiful plants in a garden is a form of practical art that requires knowledge of how to nurture plants. Art is about making—it is, in some sense, a skilled way of living. Modern definitions of art tend to refer to an exploration and expansion of perceptual awareness of the world around us.

How should we define fine art? How do we know fine art when we see it? How do we decide if Tracey Emin's famous 'unmade bed' is fine art or a con trick? Can a bridge design really qualify as a piece of fine art? In what way is a bridge different from one of Antony Gormley's famous bodies strewn around Stavanger as a broken column?

If one of the duties of art is to expand awareness, then Tracy Emin, in common with many other modern artists, uses shock tactics to focus on issues of gender, race, and sexual orientation. But shock tactics aren't the only way to expand our awareness of ourselves and of the world around us—but we do need to create a reaction. By this view art is about changing the world and getting a reaction. It is not simply about beauty but it is about taste or the validity of aesthetic excellence.

Interestingly, as we saw in Chapter 1, fine art is a historical construction of the eighteenth century. It belongs to an elite. It belongs to those who proclaim that they know what is fine art. Let's be clear that in doing so, they are saying to the rest of us that their reactions, their sensitivities and feelings, are better than ours. So they decide that Tracey Emin's unmade bed (presented as it had been when she had not got up from it for several days due to suicidal depression brought on by relationship difficulties) is fine art—many of us don't. Fine art is short-listed for a Turner Prize, chosen to be exhibited in the Tate Gallery and sold to a collector for £150,000—all of which happened to an unmade bed in 1999.

Perhaps art needs to be recaptured from that elite and made accessible to everyone. But there is a danger. This kind of argument can descend into mere relativity—'What I say is as good as what you say.' Of course everyone is entitled to a view—that must be so in any democracy. But are all views of equal quality? Definitely not. If you doubt that statement, then just imagine how you might feel as you are going under the anaesthetic and your surgeon admits to you he has never had any training or experience at heart surgery.

Bridge builders deal with Mother Nature—she is an unforgiving agent. You cannot mess with her. Everything that bridge builders do—all judgement and assertion must be well supported; otherwise, you may have a disaster on your hands and people killed. One opinion is not as good as another if it is not supported by dependable evidence. That is why bridges are built by people who are experienced bridge builders. It is why young aspiring bridge builders need to study, work hard, and learn.

So how do we judge a bridge as public art?

I am going to suggest four criteria. The first concerns the intensity of your initial emotional reaction. If when you first look at a bridge your eye is drawn, you are stimulated, engaged, absorbed, then there is artistic quality. The Gateshead Millennium Bridge in the UK had that effect on me when I first saw it. I just wanted to walk on it, touch it, and photograph it. Other bridges may annoy or irritate. When I first saw the Goodwill Bridge in Brisbane, Australia (Figure 41), I was drawn to it. I had to look at it, I had to engage with it, but I didn't like it. Other bridges, like the Millau Viaduct in France (Figure 33), create such feelings of harmony and are such impressive achievements that you just want to know much more—it has the 'wow' factor. Good art is arresting. In summary my first criterion is to what extent do you agree with this statement? *When I first saw this bridge I experienced a powerful emotional reaction.*

My second criterion is about composition and harmony. A bridge in context is an exercise in composition as surely as when an artist paints a picture or prepares a sculpture. Public artists such as Anthony Gormley are no longer content to show their pieces only in galleries; they want to place them, like bridges, in public places. An example is Gormley's Angel of the North.[14] Good composition needs balance. However, while symmetry is balance, balance is more than mere symmetry. Asymmetry can be balanced if it creates a sense of interesting flow on a visual journey without rifts or abrupt changes. An ugly bridge may create interest simply through its ugliness but the feeling of disharmony makes you

FIG 41. The Goodwill Bridge, Brisbane, Australia

draw back. For example, if a bridge sits heavily on the ground or if it is cluttered and lacks a lightness of touch it will appear out of balance. An angular bridge can be ugly because it has only straight lines with clashing hard edges. Curves create softness and emotional warmth. Details that result in deterioration of the structure, e.g. rust on concrete, can soon make it look ugly and run down. Natural flow works when lines follow the flow of the internal forces which is most often the case in hanging structures. Good composition also requires a bridge to fit its context. Appropriate proportion and scale are all-important. If a bridge dominates or is too small for its location then it will not work aesthetically. The Miho Bridge, Japan (Figure 42), is a perfect example of a bridge at one in its location, as is the Salginotobel Bridge (Figure 13). In summary my second criterion is to what extent do you agree with this statement? *The bridge is in total harmony with its context.*

The third criterion is about frozen movement. Photographic snapshots of people are more interesting when people are not merely standing still with arms by their sides but are doing something natural. A sculpture is effective when there is implied move-

FIG 42. Miho Bridge, Japan

ment. For example, the muscles of Michelangelo's David impress you because you feel David is strong and athletic—but he is just a static slab of stone after all. So you become engaged with what is going on and you imagine. You see potential movement or frozen movement in a static object. Just as a home snapshot or a sculpture is more interesting if there is frozen action, so a bridge can be more interesting if it seems slightly off balance.

The Puente del Alamillo, Seville (Figure 43), looks potentially unstable but the leaning tower and the cables mutually support each other into a single whole. There is a sense of strength but also of fragility. It is this conflict that makes the bridge an example of frozen movement. You find yourself wondering what might happen. It is this worry that concerns many engineers because frozen movement may be provided at the expense of structural logic. Indeed some engineers say that the Puente del Alamillo is structurally illogical. Ian Firth, who designs large bridges, says it is fighting gravity—there is too much effort in holding up the mast. In summary my third criterion is to what extent do you agree with this statement? *The bridge gives me a sense of frozen movement.*

The fourth criterion is about clarity and unity of flow of line, shape, texture, contrast, and form. It is the very essence of what a bridge is about aesthetically. Is its form pleasing to the eye? Do you get a sense of the parts coming together to form a whole? Simplicity may be sufficient for clarity of flow but it isn't necessary. Simplicity can be plainness, lack of inappropriate ornament or pretentiousness, freedom from intricacy and numerous divisions into bits. Simplicity must be understandable but it needn't be austere. Simplicity is straightforward, truthful, and direct. It implies modesty, innocence, and purity with integrity and unity. Simplicity may be easy, natural, and primitive. Clarity of form is the essence of what the bridge is about aesthetically. The Second Severn Crossing has clear lines both in the support span continuous beams and in the cable-stayed central spans. Personally I find it bulky and awkward—many judge it less harshly.

The opposite of simplicity is clutter. It is distracting and takes our attention away from the essence of something. So by reducing clutter we can more easily focus on what is important. The reason for my discomfort with the Goodwill Bridge in Brisbane, Australia (Figure 41), is that it seems to lack a clear flow of internal forces and has unnecessary ornamentation. It is, however, a bridge that grabs attention. In summary, the fourth criterion is to what extent do you agree with this statement? *The bridge has clear form that makes sense to me.*

FIG 43. Puente del Alamillo, Seville, Spain

Santiago Calatrava designed the Puente del Alamillo, Seville (Figure 43). He is one of the few bridge designers who could claim to be producing not just beautiful bridges but also works of public art. He was born in 1951 in Benimamet, a Valencian village in Spain. At eight years old his brother enrolled him at an art school in a neighbouring village. He drew in charcoal pencil but was learning from engravers, glass craftsmen, and carvers—people who were still trained in the nineteenth-century traditions. Later his parents sent him to high school in Valencia—something he says was a mistake because it left him no time to do other things—particularly to draw. After training in art and architecture, he went on to study engineering and took a doctorate at ETH Zurich in folding structures.[15] He said when he finished engineering that he had to re-educate himself in architecture. He was influenced by Gaudi, Nervi, and Candela. He valued the Vitruvian ideas of *firmitas, utilitas,* and *venustas*—strength, function, and delight. In 1981 he opened his own office in Zurich and began winning design competitions. His rise to fame was meteoric, as he won numerous awards such as the Gold Medal of the Institution of Structural Engineers in 1992.

Mobility, Calatrava says, is implicit in the concept of strength. It is this view that led to my choice of criterion 3. He sees 'strength as like crystallized movement'. He wants movement to be explicit—to introduce the dynamics. Calatrava clearly has the ability to work as sculptor, architect, and structural engineer. Very few sculptors can make forms as big as a bridge. Each discipline has characteristics from which the others can benefit.

Calatrava works across three disciplines, and in the next chapter we turn to a subject that applies across all disciplines—risk.

6

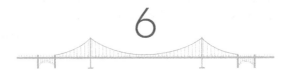

HOW SAFE IS SAFE ENOUGH?

Incomplete Science

'What the sea wants, the sea will have,' according to the traditional wisdom of the British and other maritime cultures. Certain fatalistic sailors of the past—and some of the present—never learned to swim. Scottish law once required fishermen to wear a gold earring, which was used to pay for funeral expenses if they were drowned and washed ashore. This kind of fatefulness is not attractive—I think if I were a sailor I would want to be able to swim. At least that way I might have some chance of survival from a shipwreck—some chance of being safe.

Safety is good—it keeps us alive. Yet, talk to most people about safety and they find it hard to stifle a yawn. Safety is not the most exciting of subjects—until you realize its importance for your own well-being.

On a construction site you are required to wear a safety helmet. Since bridges are usually built at some height off the ground, most accidents are down to gravity. Things can fall on you or you can fall off things—both from a great height. But there are many other horrific possibilities that have actually happened. For example, you could get your long hair caught in a machine and have your scalp ripped off. Accidents on construction sites are all too frequent despite the efforts of many people to improve the industry's track record.

The problem for most of us is that health and safety at work can be too cautious in our overly litigious society. Safe is unadventurous—and laws can sometimes be overprotective. 'Safe is boring,' says the typical teenager and rightly so. If we don't take risks when we are young we never will. Most of us like some sort of adrenalin rush but a few take it to extremes—why else would anyone bungee jump 192 m from the Auckland Sky Tower in New Zealand?

So we have a dilemma. We need to take risks to explore the world around us, to learn, to have some personal excitement and challenge, yet we all have our comfort zones—those things we can do without getting anxious. We have boundaries.

Bridges are no exception. Bridges have boundaries too. They are threatened by all sorts of hazards, almost all of which can be traced back to 'Mother Nature'. In this chapter we will explore those threats and how bridge builders deal with them. We will ask how safe is safe enough? We will examine the difficult idea of risk—how do we predict it and how do we manage it? How do we push back the frontiers and yet keep safe? We will wonder whether failures are inevitable.

Mother Nature requires respect—she will search out any weakness in a bridge, sooner or later. That is why, in the past, bridge builders have sometimes seen her as an adversary—someone to be controlled. Any sign of weakness and she will inevitably find it out and you and your bridge will be in trouble. Now we have realized that we must learn to work sustainably in harmony with nature—but it is a working relationship that demands total regard.

Safe is not boring for bridge builders—it is the challenge at the heart of their very being. The possibilities are many. At one extreme is a total collapse from a variety of causes including even terrorist attack. At the other is inconvenient behaviour, including minor wobbles and repairable damage. Bridge engineers must identify and manage the risks of all these possibilities becoming the reality. To do this they need to be imaginative and creative with the foresight to think of everything that might happen to the bridge and have contingency plans in place. They know that unintended and unwanted things may happen—so the motto is 'be prepared for the unexpected'.

One recent example of how engineers must leave their comfort zone was the rescue of the Leaning Tower of Pisa. Obviously the tower is not a bridge but nevertheless the story is instructive.

Imagine being asked to take responsibility for doing something to such an important and world-famous tourist attraction— you can just see the newspaper headlines if you get it wrong. Professor John Burland of Imperial College, London, was asked just that question and to respond to it he definitely had to leave his own comfort zone. Once on a flight back from Pisa he heard a not-so-*sotto-voce* comment from the back seat; 'The things that some people spend their money on!' John was carrying a large model of the Leaning Tower that looked like 'tourist tat'. He bought it to play a starring role in a demonstration during a lecture at the Royal Institution.

John spent more than 12 years helping to prevent Pisa's medieval masterpiece from becoming a pile of rubble. The challenge was to straighten the Leaning Tower—enough to stop it from falling over without making it so vertical that visitors feel cheated. In 1990 the Tower's movement, of roughly 1 mm per year, had brought it to the brink of collapse. It had simply become unsafe for tourists. The tower was built on very soft sediments— a bit like trying to build a tower of toy bricks on a soft carpet. The team set up by the Italian government that John had joined identified two distinct risks. First, the tower could collapse because the fragile masonry of the tower failed. Second, the ground around the foundations could break up, with the same consequences.

In 1993 a lead counterweight of about 600 tonnes was placed on the north side of the tower's base in order to stop the southward rotation. John's solution was to extract some soil from two layers of the earth beneath the tower—the top layer of sandy soil and the second of marine clay. As the soil was removed, the ground compressed and the clay firmed, giving a stronger foundation. Special drills, 200 mm in diameter, were used—designed so that there's no disturbance to the ground on the way in. When the drill was pulled back out a bit, the cavity which was left closed very gently. As a result, the ground above subsided a little bit and took the tower with it.

For the first few tense weeks of drilling in February 1999 nothing happened. Then the tower started to respond. Each time they extracted between 15 and 20 litres of soil and literally steered the tower. If it went a little bit to the east, they took a little bit of soil out on the west side and it came west. By the time work finished in June 2001, the tower had been returned to the position it was in 1838, leaning just over 4 m off centre. The tower was reopened to the public in December 2001.

John Burland's story is an example of the clever way in which we can use our modern scientific understanding to control something we previously thought was uncontrollable. It shows that we can manage risk by steering a clear process to success based on evidence from many sources.

Our collective scientific understanding and technical abilities are now very impressive. Our successes, from the Pisa tower, space exploration, through wireless communications to life-saving medical treatments are dramatic. So much so that sometimes we get carried away by them. Our high expectations are sometimes not underpinned by a firm grasp of the risks. Failure can then seem like negligence—someone must be to blame.

Unfortunately, and all too often, someone is to blame—but not always. It is important to remember that there are still many things we cannot control. We must trust to fate. Perhaps that is why the weather is such a constant topic of conversation in the UK.

Being able to control something is not the same as being able to influence it. We are now starting to understand the way we have influenced the climate and to realize the enormity of the threat of global warming. As we do so we are beginning to identify what we need to do to reduce its impact—including our new and existing bridges.

The UK, where I live, is fortunate not to suffer some of the very extreme events of weather such as hurricanes, tornadoes, earthquakes, droughts, and monsoons, commonplace in many parts of the world. But throughout the world every bridge owner must think about contingency plans.

So how do bridges remain safe? In Chapter 3 we saw that internal forces in each degree of freedom of a bridge are of two kinds—applied and resisting forces. You'll recall that applied forces are those caused by the external loads on the bridge, whereas the resisting forces result from the various ways in which the bridge is strong. The bridge is safe as long as the resisting forces are bigger than the applied forces.

We know that there are three ways in which a material can be strong—tension, compression, and shear. Our tug-of-war rope has two important limits to its strength in tension—which we usually express as stresses. They are the yield stress (which is close to but different from the elastic limit mentioned in Chapter 4) and the ultimate or breaking stress. If we stress the rope above the elastic limit then the material will have some permanent deformation. At the yield stress the material begins to flow plastically, which means that the strain increases with only a small increase in stress.

In Chapter 4 we saw that a strut has a compressive strength also usually expressed as a stress. Again there are two important limits. For very short struts the first limit is called, rather descriptively, a squashing stress, with two sublimit values, the yield and ultimate stresses (as in tension but squashing). For longer struts the limit is a buckling stress, which is quite variable. It depends crucially on the geometry of the strut, the way it is held at its ends, and most importantly on its lack of straightness.

There are also two common ways of expressing the limits to the strength of a material in shear. These are a safe maximum value at a point and a safe maximum average value over a stated area. So the maximum limit on a shear force on an I-beam may be when the local shear stress at the centre of its web is at a maximum. More commonly the shear force limit may be found by multiplying the whole area of the web by a maximum average value of shear stress.

The safe values of other ways of expressing the strength of a bridge are derived from these three basic ways of being strong.

I will now illustrate how this works for bending moments in the beams that we looked at in Chapter 3.

You'll recall that a bending moment is created by the turning effect of tensile and compressive stresses in a beam.

If the turning effect is caused by an applied load then we get an applied bending moment as described in Chapter 3.

Now we want to know the turning effect that the beam can resist. This comes from the internal stresses in the beam when they are at their largest. We call it the bending moment of resistance. We need the bending moment of resistance to be greater than the applied bending moment.

So let's now find the bending moment of resistance.

All of the internal stresses are unlikely to be at their largest at the same time.

So let's look again at Figures 18b and 18c, re drawn as Figures 44a and 44c. However, this time some of the forces and distances have been labelled. A cross section of the beam, rectangular with a width b and depth d, has been included in Figure 44.

The maximum stresses in Figure 44a are at the top and bottom of the cross section. In this beam they are equal and labelled s. So, for example, s might be a stress of 200 newtons per square millimetre (N/mm^2).

You will recall from Chapter 3 that the yield stress is the largest stress that a material can take in tension or in compression before the strain grows rapidly with no increase in stress.

So the stress s is at its largest when it is equal to the yield stress of the material from which the beam is made. The value of s at the bottom of the beam is the yield point in tension, and the other at the top is the value of the yield point in compression. For

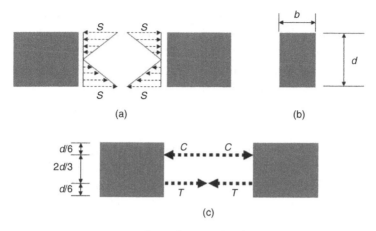

FIG 44. Internal stresses in a beam

mild steel these are both 245 N/mm². We can now start to find the resisting bending moment from Figure 44c.

You'll recall from Chapter 3 that the tensile force T on each face of the cut in Figure 44c is the sum of all of the tensile stresses in the triangular block in the bottom half of Figure 44a. Likewise the compressive force C on each face of the cut is the sum of all of the compressive stresses in the triangular block in the top half of the beam.

T and C together have a turning effect—the resisting moment RM, which can be calculated as

$$RM = (s \times b \times d^2/6).$$

This formula tells us that the resisting bending moment (RM) of a beam depends on yield strength s of the material from which it is made and the shape and size of its cross section.[1] In particular for our rectangular beam the resisting moment depends directly on the width b, and the square of the depth d.

For other shapes we'll get different answers but they will still depend purely on the geometry of the cross section.

So we have a safe limit for the bending moments in the simply supported beam of Figure 19. In Chapter 3 we saw that the applied moment is $(W \times L/4)$. We'll call that AM.

Remember there are many limit states—we are just looking at one of them. The beam will be safe in this particular limit state if RM is greater than AM or expressed in full if $s \times b \times d^2/6$ is greater than $WL/4$.

Bridge builders define a safety factor, which is the resisting moment divided by the applied moment or (RM/AM).

Typically this may have values from 1.1 to 2 so the reserve of strength might be just 10% when we know the loads and strengths quite accurately and might be 100% when we don't. However, this safety factor is quite a crude way of dealing with risk—we'll look at risk in more detail in a moment.

Before we do let's just summarize what we now know. First there are three ways in which a material can be strong—tension, compression, and shear. Second, there are a number of different limits for these three stresses. Examples are elastic limits, yield, and ultimate stress values. These limit states define various boundaries beyond which we mustn't go if the bridge is to remain safe. Third, the strength in all of our limit states still depends on the three ways in which a material is strong—although often with other complicating factors. We have demonstrated it for the bending moment in a simply supported beam with a rectangular cross section when the maximum stress is a yield stress.

I will mention three further examples to illustrate the point and some of the complicating factors. They are torsion, vibrations, and fatigue. A limit state in torsion is the twisting of a beam, as distinct from the bending of a beam. The limit states are derived very similarly to bending but are quite a bit more complex and depend on limiting shear stresses as well.

We looked briefly at bridge vibrations in Chapter 5. They depend crucially on the fundamental frequencies of the bridge. These in turn depend on mass, damping, and stiffness—all characteristics of the materials and geometry of the bridge. However, subtle interactions between the forcing frequencies and the modes of vibration, as in the flutter of the Tacoma Narrows Bridge, make this a complex matter even for specialists.

Fatigue behaviour, again as we saw in Chapter 5, is not straightforward. As well as the bridge characteristics it depends on the number of cycles of the applied load, the brittleness of the material, which is affected, for example, by the way welds have been laid down, the temperature, and many other detailed factors. Limit states in fatigue can be highly uncertain.

However, in every case bridge builders must make sure that the applied forces are less than the resisting forces by a suitable margin. In other words the risk of the applied forces becoming greater than the resisting forces is acceptably small.

So how could these risks turn into reality? There are three ways and each is complex. The bridge builders could just underestimate the size of the likely applied forces on the bridge or the likely strength in a limit state as we have seen earlier. Or their understanding of the way the bridge works could be faulty. They have simply misunderstood the flow of applied internal forces and how they will be resisted. Or there could be a human error.

At one extreme this could be a simple mistake. At the other end it could be a complex organizational issue such as a lack of 'joined-up' thinking, which we will look at in detail in Chapter 7. In between there could be terrorist sabotage or even criminal activity.

Let's look at each in turn, starting with the sizes of the likely applied loads and strengths.

Earlier in the chapter we defined a safety factor. Now we need to consider the nature of the loads and the statistical chance that any limit state will be exceeded during the lifetime of a bridge. The only mathematical theory of chance is probability theory. Unfortunately, it is totally inadequate for estimating the risk of bridge failure—later we will see why.

But let's start thinking about the risk of bridge failure by looking again at the likely applied loads. Loads are actions applied to a bridge. There are two types—permanent and variable over time. The permanent loads are called dead loads and include the self-weight of the bridge and all of its fixtures and fittings. Variable loads are many and various—some are man-made, some are natural. Man-made loads are called live or superimposed loads. So the questions the bridge builders must ask are 'What is the biggest truck or train that might cross the bridge? How many trucks or trains will be on the bridge at any one time?' If the bridge is for pedestrians only, then the questions are 'How many people of what weight will be on the bridge at any one time?' Natural loads are the environmental loads, for example wind, flooding, and earthquakes and settlement of foundations. Changes in temperature along the length and through the depth of a bridge must be accommodated. Other real threats include the impacts of collisions, scour of foundations, snow, fire, and explosions. The bridge builder must estimate the likely size of all of these threats to his bridge and make sure that the bridge is strong enough to resist them.

The easiest load to estimate is the dead load but it is a chicken-and- egg problem. Bridge builders usually start their first draft design by assuming a typical figure for the dead load which is distributed uniformly along the bridge (shown as the wiggly line in Figure 35). They then calculate the dead weight of their proposed bridge and compare it with the assumed figure. If the difference is unacceptably big then they must reanalyse the bridge and adjust the design. So the better the first guess, the less reworking they must do. Clearly as they know the self-weights of all of the materials the dead weight can be estimated fairly accurately. Of course there is a margin of error, so safety factors for dead load only are usually around 1.1.

The live loads are more uncertain. They can be estimated from the kinds of traffic expected on the bridge. But they can change dramatically. The Clifton Suspension Bridge (Chapter 5) was designed to take horse-drawn carriages— now it must take modern vehicles.[2]

Bridge builders before the twentieth century, such as Brunel, Telford, and Roebling, had some freedom to decide what the live loads might be. Modern builders don't. They must follow regulatory rules set by governments to protect the public from inconsistent decision-making. Every country has rules and regulations that stipulate how the bridge builders should decide on loads and strengths.

The first English construction regulations, written in 1189, stipulated rules about party walls, ancient lights, and foul cess pits. In 1620 James I proclaimed the thickness of walls.

A standard brick was defined in 1625. The first British Standards were written at the beginning of the twentieth century. In 1922 the first British Standard for steel girder bridges specified maximum permissible stresses for the three ways of being strong in tension, compression, and shear.

Now regulations cover almost all aspects of bridge design. In the UK the British Standards are being superseded by Eurocodes, which are technical regulations for all countries of the European Union. Similar regulations based on LRFD (load and resistance factor design) apply in the USA and other parts of the world. Typical examples are the requirements for different live loads for highway, railway, and pedestrian bridges. The highway loads usually consist of an equivalent distributed load which represents a traffic queue with particular wheel loads to represent very heavy trucks. These can weigh in the region of 180 tonnes.

Regulations also cover other issues such as the impact factors for collision loads (Chapter 5). Derailed trains, out of control trucks, even ships can and do sometimes collide with bridge piers. For example, in 1975 a bulk carrier brought down three of the twenty-two spans of the Tasman Bridge in Tasmania. Seven seamen were killed, four cars drove off the edge killing five people, and the ship sank. Bridge builders usually deal with this kind of dynamic load by multiplying the weight of the train, truck, or ship by an impact factor to find a larger equivalent static load. It seems that they got it wrong for the Tasman Bridge.

The wind and earthquake loading on a bridge is complex and the subject of much research especially for long-span hanging bridges. Clearly wind load depends on location. It depends not only on latitude and longitude but also on the local position, whether in an exposed situation, such as a deep gorge, and the roughness of the surrounding terrain. Modern bridge builders again have guidance through regulatory rules.

For smaller bridges when the likely interaction between the wind and the movement of the bridge is small then the wind is modelled by a sideways static distributed load considered equivalent to the steady pressure of the wind over a considerable period such as 1 minute.

For the big bridges, after the failure of Tacoma Narrows Bridge, a model of the bridge will be tested in a wind tunnel. Engineers look for interactions between the flow of wind and gusts with possible modes of vibration. Wind tunnels are laboratory facilities where air is blown past a model of the bridge. The response of the model is monitored and measured together with wind pressures around the bridge. This is called an aerodynamic analysis, and the results are input into a finite element model of the bridge to find the flow of internal forces.

Likewise if a bridge is to be built in a seismically active zone the bridge will be analysed by simulating the bridges response to an earthquake.

In some cases a model of the bridge may be tested on an 'earthquake-shaking table'. The idea here is similar to a wind tunnel in that a model of the bridge is built on a table which can be moved in all six degrees of freedom. You'll recall that those are side to side (x), back to front (y), up and down (z), with rotations in each plane (x, y), (x, z), and (y, z)—see Figure 45.

Other threats to a bridge are fire, explosions, changes in temperature, snow, flooding, and scour and settlement of the foundations. Fire and explosions can be accidental (as at Britannia Bridge, Chapter 3) or deliberate acts of sabotage. Temperature changes, whether daily, seasonal, or long term, cause the material

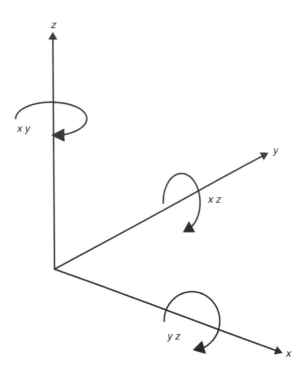

FIG 45. Coordinate axes

of the bridge to change in size. If those changes are restrained in any way by the rest of the bridge or its supports then internal forces will be set up. Those forces will have to be resisted by the materials of which the bridge is made. If the stresses are unrecognized and get too high then the bridge may be damaged.

It's interesting in passing that one of the upsides of a statically determinate bridge is that all movements from temperature changes can be accommodated without creating extra internal forces. The downside is that that very attribute makes them vulnerable and lacking in robustness against small amounts of damage.

So far we have assumed that the strength of a material whether in tension, compression, or shear is a constant value that we can determine. Unfortunately this is not so—there are statistical variations.

If you were to measure the length of a room with a tape measure several times, you would get several slightly different answers. In the same way if we take many pieces of our tug-of-war rope and pull them, in a special machine, until they break, we will find that the breaking stress each time will be slightly different. In such cases the results often (but not always) follow the bell shapes shown in Figure 46.

How do we interpret the bell shape and what value should we use for the limiting strength of the rope?

The bell shape expresses the chance or probability that the size of a load or a strength value lies in a particular range of values. Three main features describe it. The first is the mathematical formula for the curve. The second is the average or mean value where the curve is a maximum.[3] The third is a measure of spread or the width of the bell shape, which is called the standard deviation. The curve is drawn so that the area underneath it is equal to 1.

We could use the average strength since it is most representative. However, that might be dangerous since the rope used for the bridge might easily be below average and we will have less strength than we thought. So we should use a lower than average strength. This reduces the chance that our actual rope will be weaker than we have assumed but it doesn't eliminate it—there is always a chance that the rope is weaker.

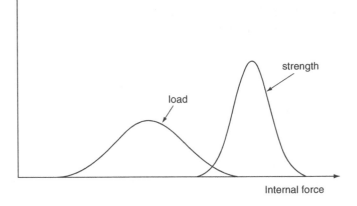

FIG 46. Distribution of load versus strength

So when checking the safety of his bridge the bridge builder usually assumes that the strength of his bridge, as designed, is very low because in that way he is assuming the worst. He therefore assumes a value for the strength of his bridge which is at the lowest (left-hand) end of the tail of the bell-shaped curve for strength shown in Figure 46.

The variations in tension test results on the tug-of-war rope will be quite large but for a more homogeneous material like steel they will be relatively small. However, if you do a similar test on slender steel struts you will find a much larger spread. This is because the way the test is set up and performed is critical—in particular the initial imperfections in a strut are so very important.

Just as the strength of the bridge varies, so the load will vary too. Unfortunately available data are sparse, making it difficult to establish a dependable bell-shaped curve.

So the bridge builder's problem is that he has two types of forces which could take on a number of possible values just due to random variations. The first is all of the internal applied loads resulting from the external applied loads. The second is all of the internal strengths which must resist those loads.

He needs to be sure that, in every case, the strength is always greater than the load. So what does he do? He assumes a very low strength (as we said earlier, at the left-hand end of the bell-shaped strength curve in Figure 46) and a very high load (at the right-hand upper end of the load curve). He then checks that, in this case, the assumed strength is greater than the assumed load by a suitable margin—called a safety factor.

However, what he really wants to know is the chance that the strength might happen to be less than the load at some time in the future. This is the chance of failure. If it is too high then he will need to redesign his bridge to make it stronger.

So the event he is interested in occurs in the overlap of the two tails of the two curves. He needs to know two probabilities. The first is the probability that an internal force caused by an applied load has a particular value, say F. The second is the probability that the strength is less than F. The probability of failure for F is the product of those probabilities. But F can vary so he then must repeat the calculation over all possible values of F and then sum them all up.

Although this probability of failure is reasonably easy to describe, it is by no means easy to calculate. Engineers use the mathematics of probability theory but the maths is complicated. Of course the overlap must be very small, so the probabilities of failure may be in the region of 1 in 10 million. Unfortunately for engineers who must analyse these data the answers are very sensitive to all sorts of assumptions. The results therefore are only indicative, not particularly useful, and potentially very misleading.

Nevertheless engineers have developed algorithms to calculate probabilities of failure. The results are only indicative because they are not statistical chances. The reasons are simple— I will mention just two. The first, we have already said, is that the results are very sensitive to difficult assumptions about the nature of the curve—the bell shape is only one of the many possibilities. The second is that there are many reasons why bridges fail that aren't included in the calculation. This makes the calculated figures potentially misleading. Usually they are called 'notional probabilities of failure' to reflect this deficiency but quite often the word 'notional' gets dropped. These partial results then have the appearance of real statistics.

The second way that a bridge could fail is when our understanding of the internal forces is faulty. Nowadays that understanding is based on science. Bridge building is a magnificent example of the practical and everyday use of science.

Naturally the science applied can only be as good as the science known and the assumptions made. So the way science is used is a process that has changed through time. The earliest applications were, to the modern mind, simply mathematical rules. The ancient Egyptians used rules to estimate the area of their fields though their understanding would have been different from ours. The Greeks, through Pythagoras, Euclid, and many others began to formulate the basics of the mathematics we use today.

The Romans were great bridge builders. There was no need for rules about arches since they were circular but rules were developed for other structures. In the first century AD, Vitruvius, a Roman architect, wrote that the columns of basilicas should be 'as high as the side isles are broad; an isle should be limited to one-third of the breadth which the open space in the middle is to have.'

Other rules were more like recipes:

> The foundations of these works should be dug out of the solid
> ground, if it can be found, and carried down to solid ground as far
> as the magnitude of the work shall seem to require, and the
> whole sub-structure should be as solid as it can possibly be laid.
> Above ground, let walls be laid under columns, thicker by half
> than the columns are to be, so that the lower may be stronger
> than the higher.[4]

In the mid-nineteenth century similar rules were still being
used, this one for masonry arches is attributed to Thomas Telford:

> If we divide the span of an arch into four equal parts and add the
> weight of one of the middle parts one sixth of its difference, from
> the weight of one of its extreme parts, we shall have a reduced
> weight, which will be to the lateral thrust as the height of the arch
> to half the span. ... In order that an arch may stand without
> friction or cohesion, a curve of equilibrium proportional to all the
> surfaces of the joints must be capable of being drawn within the
> substance of the blocks.[5]

Rules are still used even today. They are needed because the
science is either too complex to be used effectively or it is
seriously incomplete. For example, rules determine the spacing
of bolt holes in a steel plate. Rules help designers proportion
bolted and welded steel joints. These details can be very difficult
to analyse theoretically because there are many sharp changes or
discontinuities in the flow of the internal stresses.

By the middle of the twentieth century a new 'plastic theory'
was developed which enabled bridge builders to calculate the
collapse load of a bridge. You may recall that in Chapter 4 I said
that some materials, like steel, behave in an elastic way up to an

elastic limit then the material goes 'plastic'. This means that the strain increases without any increase in stress. But it happens only at certain places where the stresses are the greatest. As the strains flow, then the stresses change and the internal forces get redistributed. They remain balanced with the external forces but balanced in a different format.

We can see how this works by re-examining the stresses at the cut in the simply supported beam of Figure 18 and Figure 44. Let's imagine the process as the load gets larger and the applied bending moment increases. We know that the largest stresses are at the top and bottom in Figure 44a and so they will be the first to yield and the strains to go plastic. I have drawn the beam yet again in Figure 47a. As more and more bending moment is applied these yield stresses on the outside cannot get any larger but the strains can and do. As a consequence the stresses lower down within the beam get larger and eventually they go plastic too as in Figure 47b. Indeed eventually all of the stresses right through the cross section reach the yield stress and the whole cross section becomes totally plastic in Figure 47c. So instead of the triangular distribution of Figure 44b we get a rectangular distribution for tension and for compression with all stresses at the yield point.

When this happens the beam has no stiffness at the cross section—the strains just go on increasing. The situation is rather like a rusty door hinge as the beam rotates but also resists a bending moment. Now the whole beam deforms as the hinge at the centre rotates and the beam collapses as in Figure 47d.

It takes only one plastic hinge to collapse a simply supported beam because the ends of the beam are free to rotate and you effectively have three hinges in a row. If the ends were fixed and

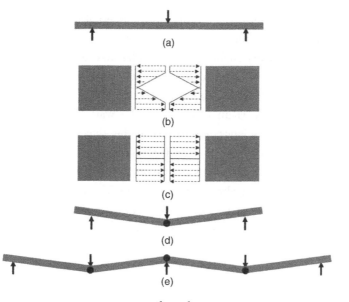

FIG 47. Plastic hinges

encastré (Chapter 3) then plastic hinges would have to form at the ends before it would collapse. The two-span beam of Figure 20 has been redrawn as Figure 47d. You can see that three hinges are needed for it to become a mechanism and to collapse. In general a statically indeterminate bridge needs a sufficient number of hinges to form a mechanism—rather like the arch of Figure 10f.

In summary this new theory of plastic collapse brought two separate benefits. First, the redistribution of stress within a hinge increased the estimated strength of the beam. Second, the process by which further hinges were formed also increased the estimated strength of the bridge as the internal forces are redistributed throughout the entire structure.

The new theory enabled bridge builders to examine a structure as the stresses go beyond the yield stress and up to the stage when the whole bridge collapses.

Plastic theory is an example of the way in which bridge builders apply science. But in doing this they are not merely applied scientists because bridge building requires so much more. Science is always necessarily incomplete when applied in practice—there are always big unknowns. The application of the theory depends crucially on context. Consequently professional judgement based on experience is always required.

That is why we tend to use the word 'model' to describe a scientific representation of a bridge. 'Model' has at least two uses. The first is familiar—a model is a simplified physical representation of something. A toy model car for example has the shape and appearance of a car but usually without an engine or detailed working parts. Another is a physical model of a bridge used to show clients and public what the bridge will look like. But models of bridges are made for scientific use too. The erection process may be tested in a wind tunnel or a particular part of a bridge may be pulled, pushed, bent, or twisted in a laboratory testing rig to see how strong it is and at what load it breaks—just as Fairbairn did on a model of the box beam of the Britannia Bridge (Chapter 3).

The second use of 'model' is less familiar perhaps. A model is also a theoretical representation of something. Indeed it can be argued that all mathematics and science is a model because it is a way of representing something other than itself. It is a model of the way we understand the world around us. The meaning derives entirely from the way we interpret the model as representing the reality. So when we talk about the internal forces in a bridge we are inevitably talking about a model of the forces in the bridge. Everything in our heads, all of our understanding, is a set of patterns in the brain which model our experience and understanding of the world around us. And when faced with the reality of using the model to make something real we must recognize that all models are incomplete and can be wrong.

This facing up to the reality of making a big bridge is the crucial difference between all practical uses of science and science itself. Both medical and engineering practitioners must make sure that what they are doing will work—or at least the risk of it not working is acceptably low. Scientists test their hypotheses under controlled conditions but rarely are the consequences of failure catastrophic. A counter-example may be manned space flights but they are really engineering projects to support scientific purposes.

It is the incompleteness of the unexpected and unintended that makes the mathematics of probability theory inadequate as a theory of risk. Probability is a quantitative measure of the likelihood of an event. If we are sure something will happen the probability is one. If we are sure something will not happen the probability is zero. If we are equally unsure one way or the other then the probability is one-half. Hence the expression 50–50 (this is a percentage of course). In certain contexts this definition is fine because to find a probability we just count the possibilities. If we roll a dice then the possibilities are 1, 2, 3, 4, 5, 6 and so the probability of rolling a 3 is 1/6. The probability of rolling an even number is 1/2 because there are three even numbers out of six possible results.

So there is a basic assumption in probability theory—that we know all of the possible outcomes to an event. We call this type of problem a *closed world* problem—one in which all possible outcomes can be identified precisely.

In a moment we will consider the possible ways in which we humans make errors—here the sample space is infinitely large. We call this an *open world* problem.

Some mathematicians who are keen on using probability theory say that they can tackle open world problems by the use of so-called subjective probability. In this theory every decision is, in effect, modelled by considering an 'equivalent lottery'—a 'virtual' gambling problem. The probability measure is then interpreted as a degree of belief. The theory is based on conditional probabilities. This just means that whenever we state a probability we state it given that another event or set of events has already happened. An algorithm, called Bayes' Theorem, is used to continually update the degrees of belief based on new information.

The reason why this theorem is rarely used in bridge building is because the assumptions are too gross. For example, no contradiction and no ambiguity are allowed, so much of the information that bridge builders need to sort out to decide how to act cannot be included.[6] Classical probability theory is not a good model of the decisions that bridge builders make take.

However, a difficult question now looms large. If models are indeed representations and are necessarily approximate, then we need to know 'How good is the model?'

Clearly some models will be better than others. The emphasis in bridge building is not the scientific one of 'Is the model true?' Rather it is the practical one of 'To what degree can we depend on the model to make practical decisions?' Bridge builders are not concerned with the truth of a model but with how good it is. They need to know the extent to which it is dependable for its purpose. Truth is sufficient but not necessary for that dependability.

Engineers create scientific models and use them to predict behaviour. Of course things that are not anticipated may happen. It follows that we may not be able to predict the risks to a bridge accurately. The results of models do not pretend to represent the future as it will be—but rather as it might be. The emphasis is not on accurate prediction (though that is important and useful if it can be done) but rather on designing and building something that will do the job required of it.

The aim is to build a bridge that will be fit for its purpose. Any predictions and indeed all evidence, about past, present, or future, is variously trustworthy, depending on how it is interpreted from a good understanding of the models we use set in their separate contexts. So when the bridge is built, and is being used, then those responsible for it must be ready for whatever happens. Bridge builders need imaginative and creative foresight to think of everything that might happen to the bridge and have contingency plans in place. They know that unintended and unwanted things may happen—so the motto is 'be prepared for the unexpected'.

To recap, we have now looked at two ways in which a bridge could fail—random overload or understrength and uncertainty in our models.

Now let's look at the third way—'human error'. You often hear reports about accidents being blamed on human error. This is about as helpful as saying gravity caused a vase to fall off a shelf.

We must unpick its meaning very carefully and I can only skirt around some of the issues here.

Human error can vary from a simple mistake to a complex organizational mix-up. How do bridge builders cope with it? The main thing they do right from the start is to recognize that individual people have the potential to do the most unexpected things. But more than that—when people get together in organizations all sorts of unintended consequences can emerge.

There are three kinds of individual human error—excluding, as I will, criminal activity. First, slips and lapses occur when you intend to do the right thing but it turns out faulty. You accidentally spill some coffee and stain the book you are reading or you mix up two digits in a telephone number. Second, mistakes occur when your plan is faulty. The young graduate engineer made a mistake at the Second Narrows Bridge (Chapter 4) because his calculation was based on a wrong assumption. Third, violations occur when a deliberate but not necessarily blameworthy deviation is made—a non-conformance. You do this when you ever-so-slightly exceed the speed limit in your car.

Bridge builders can, by careful checking, reduce the chances of slips and lapses and violations affecting the safety of bridges. Mistakes are more difficult. We all make mistakes—that's just a fact of life we must recognize. Of course the key to success is to try to learn from them and to not repeat them. To do that we must understand and then we must change. The wobbly bridge builders of Chapter 1 understood and changed.

To learn from our mistakes we need to reflect on them, learn the lessons, and try to spot what we need to do to avoid that mistake again. But we will only get the benefit when we use what we have learned. This is often much more difficult than we might expect—especially within organizations.

Barry Turner was a sociologist who, for a time, studied engineering. He spent his academic professional life investigating man-made disasters. He and I with psychologist Nick Pidgeon worked together for nearly 15 years before Barry's untimely death in 1996. Barry believed that there are typically five stages in the development of a man-made failure. First there is a set of initial beliefs that people hold about a situation—and they undermine it. The initial partial thinking about the wobbly bridge is an example. It might just be a seemingly small matter that people have 'got away with' for a long time—something that has become part of the culture in the sense it is an accepted part of the way things are done. For example, it may emerge after a car crash that people routinely ignore the speed limit in a particular restricted zone. Perhaps after a bridge collapse it emerges that mathematical calculations were often not being checked as carefully as they should have been because of time pressures.

The second stage in a failure, according to Turner, is that the conditions for failure incubate over time. In this stage the potential for unintended and unwanted consequences of human actions develop through multiple factors. For example, events may be unnoticed or misunderstood because of wrong assumptions about their significance. That is exactly what happened in the Westgate Bridge collapse when the bolts were undone to straighten out the buckles on a previous span. In Chapter 7 we will see how important this stage was in the collapse of the Minneapolis I-35 bridge collapse mentioned in Chapter 4. There might also be uncertainty about how to handle formal violations of safety regulations. People may spot things that worry them but minimize the emerging dangers because they can't believe it might lead to failure. It is in this second stage that we can learn a great deal about how to prevent future similar tragedies.

The third stage is the first very obvious realization to all involved that something has gone seriously wrong. There is a trigger event that results in the actual physical failure—the 'straw that breaks the camel's back'. It is the last event in a long incubating story leading up to the drama of the bridge collapse.

The fourth stage in a failure is the reality of failure and the rescue. Television pictures of buildings after an earthquake or a tsunami are stark reminders of the harrowing events that take place during this stage. There are many lessons to be learned about how to deal with the physical damage and how to improve rescue techniques with modern equipment that can locate people buried under tons of rubble.

The final stage is perhaps the longest one of all—cultural readjustment which may takes years or even decades. The pain felt by the families and friends of people killed takes a long recovery time.

So how do bridge builders cope with human folly? In brief they use their specialist knowledge and experience to plan very carefully, monitor everything that happens, and change what they do to meet their objectives. So they must become very accomplished at managing risk.

In everyday talk risk and hazard are more or less the same. However, to bridge builders and risk managers they have very different meanings. A hazard is a state of affairs with a potential for harm. For example, a trailing wire that someone could trip over, a patch of ice on the road. Wind load is a potential for damage to property, climate change is potential for damage to the environment. Hazards exist in the past, present, and future.

Risk, on the other hand, is the chance or likelihood of a specific hazard actually coming about in the future. Risk is about the future. What is the chance of some state of affairs happening at some time in the future combined with the consequences that will follow?

Hazards and risks can only be understood in context. To identify, understand, monitor, and change them we need dependable evidence. Evidence is information that helps us to come to a conclusion. It is the basis or reason for us to believe something—though we must hastily agree that, as individuals, we can quite easily believe something without evidence.

Dependable evidence comes from scientific or mathematical reasoning, from opinions expressed by suitably qualified people, through to legal testimonies in a court of law. Again I must hastily point out that I am not denying the right for anyone to hold an opinion. But I am saying that when all we have is someone's opinion then we must consider carefully the qualifications, knowledge, and experience of the person giving the opinion. If you are in any doubt on that, imagine having your teeth drilled by someone who is not a qualified dentist or heart surgery performed by someone who has no medical knowledge. It follows that bridges should only be built only by people who are qualified to do so.

In other words bridge builders know that any information they work with has a 'pedigree' which they must feel is sufficiently dependable for their purpose.

One important characteristic to be examined in assessing the pedigree of evidence is uncertainty. Put simply, uncertainty is a lack of information. Information may be incomplete, fragmented, unreliable, vague, contradictory, or generally deficient in some other way. For example, partial or 'silo' thinking with inadequate processes and a lack of 'joining up' do typify technical and organizational failures (see Chapter 7).

We can classify uncertainty into three kinds which I will call FIR—fuzziness, incompleteness, and randomness. We'll now examine these three, see their emergent properties, and their relevance to bridges.

I'll take them in reverse order. In brief, randomness is a lack of any pattern in some information. A lottery draw of 3, 17, 21, 35, 40, 46 seems clearly random since there is no obvious pattern in the numbers. If the draw was 2, 4, 6, 8, 10, 12 you could be forgiven if you said this wasn't a random sequence. However, if the process of selecting the numbers was such that the outcome is random then any sequence of six numbers is equally likely. That is what being a random process means.

The mathematics of statistics looks for patterns in data over populations, i.e. the number of items or people from which samples can be taken. Statistics requires a mathematics of randomness.

Earlier we said that the parameters describing the strength of our tug-of-war rope or a strut or beam vary when we try to measure them. One way to build a mathematical model of that variation is to treat the variation as randomness and look for patterns over populations of data.

The second kind of uncertainty is incompleteness. It is important but often forgotten and neglected. Indeed it was ridiculed when Donald Rumsfeld, the one-time US Secretary of State for Defence, famously said, 'There are known knowns—these are things we know we know. There are known unknowns … these are the things we do not know. But there are also unknown unknowns; these are the things we don't know we don't know.'

But Rumsfeld was right. We can see this if we examine how we learn. As a small child there are many things that you didn't know existed—like Newton's Laws. They may be known by others but not by you—you are completely unaware—to you it was an unknown unknown—perhaps it still is. The people who begin to study physics eventually learn about Newton's Laws, and so for those people Newton's Laws become a known known. But perhaps, like most bridge builders, they stop studying physics before getting to Einstein's Relativity Theory. They know it exists but that's all—it is a known unknown. We all have lots of these.

To a bridge builder the really interesting incompleteness, however, is the unknown unknowns that not just you as an individual don't know about but that nobody anywhere knows. We had an example in Chapter 3 on the Dee Bridge. In 1849 no one knew of the phenomenon of lateral torsional buckling of beams—it was a collectively unknown unknown. There seems to be only one way to deal with unknown unknowns and that is to tread very carefully especially when innovating. The best advice seems to be to monitor closely what is happening and be ready to take the necessary actions to prevent disaster if something unexpected turns up.

The last category of uncertainty is fuzziness. This is simply vagueness of definition. If I say that the internal stress in a beam was *high* when the loads are *low* then I convey information—but it is imprecise in two ways. The first is the meaning of what constitutes a *high* stress or a *low* load. The second is the nature of the relationship between them. We know that when one is high then the other is low—the statement has meaning but it is imprecise.

Conflict, contradiction, or disagreement occurs when incomplete and fuzzy information clashes. Ambiguity arises when we aren't sure what something means because the information is imprecise and incomplete and there is room for different interpretations. These characteristics emerge from FIR—they are an inherent part of the richness of natural language including poetry and literature, with its ambiguity and layers of meaning, as we struggle to express our inner feelings.

Of course the best way to reduce risk is to remove as much uncertainty as possible and to communicate as precisely as we are capable. That is why mathematics is the language of science—it is one of the most precise ways we have of expressing what we mean.

Data about practical activity are often sparse and therefore uncertain in all three FIR ways. Bridge builders must use their experience to make sense of it. So when we have lots of data, life should be easy. Unfortunately often it isn't. That's because there are so many ways to present the same information, it is easy to be confused as people use data to support a point of view. This a particular issue in statistics—we all know the adage—lies, damn lies, and statistics.

Statistical data are thrown at us every day. Bridge builders get it too. We need to sort the 'wheat from the chaff' but it's not easy. Again if you need to judge the dependability of the information then you need to understand its context so that you can see it for what it is—not what it appears to be. An excellent reference on this is Darrell Huff's book *How to Lie with Statistics*, a must-read for anyone concerned with interpreting conclusions based on statistics.[7]

Take, for example, the way risk is expressed. Three very different ways are commonly used but they can send very different messages. Imagine we sample 1000 instances of something like the quality of a weld. Each sample of the weld must pass a test and be declared a success or a failure. We do two lots of tests. In the first set 53 samples fail. In the second test 43 samples fail. How have the risks of failure altered?

From one perspective the risk has changed from 53 to 43 in 1000, so by 10 in 1000 or 1%. From another there is reduced risk of 10 in 53 or 19%. Another is that the number of tests needed to detect a change of one failure is 100. Which is the real one? The answer they are all real but convey different ideas and can be used by people as an argument to make their case more persuasive.

Another difficulty in dealing with risk is the power of anecdotes. Like the case histories I have included in this book they are stories from which you can learn a great deal because they represent what actually happened in one instance.

Unfortunately, anecdotes and case histories may be misleading about risks across a whole population of instances. Smoking causes lung cancer but you may well know of a 90-year-old man who has smoked 40 a day since he was 10 years old. It is the statistical relationship between smoking and lung cancer that is established—if you smoke, you increase your risk of ill health. If you do smoke you might be one of the lucky ones who escape unharmed—but the chances are small.

A very large part of the skill of a bridge builder comes from 'hard fought' experience in which uncertain information must be sifted and used. This type of practical judgement requires a type of rationality that transcends the application of strict logical rules and the appliance of science. It requires creative and judgemental thinking which is rational and logical but can, at the same time, create practical solutions despite the limitations of our understanding.

Practical rigour requires wise foresight to anticipate what can go wrong and put it right before the consequences are serious. Sloppy and slipshod thinking has no part in practical rigour but since there is inevitably a lack of fit between theory and practice, sensible approximations must be made and appropriate models chosen. Every practical possibility must be considered and every reasonable precaution must be taken.

Practical rigour is a rigour of the open world where uncertainty is profound. Practical rigour is therefore much richer in concept than scientific and logical rigour. In fact the latter is a subset because it is about completeness of rigour in a closed world. Scientific rigour requires selective inattention to the difficulties that it cannot yet address. Practical rigour does not have that luxury—it must include everything relevant. Practical rigour requires the management of incompleteness. Strict logical rigour is usually only possible if incompleteness is left out of the model.

At the start of the chapter I posed the question 'How safe is safe enough?' Now what can we conclude? Building bridges, like almost everything in life, is not risk free. Bridges will continue to fail but there is a lot we can do both to prevent loss of life and keep the risks acceptably low.

However, the problem is complex. The only way to get at a meaningful probability of failure is by collecting statistics over many years and over large populations of bridges. But these figures apply to all of the bridges in general but none in particular.

Despite our scientific successes prediction about any one specific bridge is deeply uncertain. Bridge builders manage risks not by making a scientific calculation of some mysterious number such as a probability of failure, because they know that is insufficient and applies only in a specific context. Rather they must be clear about purpose and objectives of what they are doing, monitor and decide on the basis of evidence. They must steer processes to success and manage the risks in a practically rigorous way. We'll look at these processes in more depth in the next chapter.

There is another worry deep within these processes. The risk of two aeroplanes flying into the World Trade Centre in New York City on September 11th was so low on September 10th as to be practically zero. This is called a low-chance, high-consequence event. The WTC was vulnerable to damage that caused progressive collapse. Vulnerability is an emergent property of the lack of joined-up thinking which we now turn to in the next chapter.

BRIDGES BUILT BY PEOPLE
FOR PEOPLE

Processes for Joined-up Thinking

Bridges are built by people for people. Bridge builders must work with and through others—they must work in teams. Teams work effectively when relationships are good. Relationships connect people—as do bridges—so I will call them people bridges.

As individuals we build a people bridge whenever we make a new friend, or work with a new colleague. We maintain people bridges through our social lives. At home, at work, or in a relationship some links may be relatively straightforward—making an acquaintance at a party. Others may be part of a complex situation—such as deciding to change a job or whether to get married. The most difficult are the relationships between companies, nations, and racial and religious communities. When people bridges are neglected, relationships deteriorate.

So bridge builders must create people bridges in order to create physical bridges. Just as the book of physical bridges has many levels, from letters, words, sentences, paragraphs, and chapters, so does the book of people bridges—but the book is very much more complex.

BRIDGES BUILT BY PEOPLE FOR PEOPLE

The bridge metaphor is deeply embedded in our thinking. Perhaps the most profound example is the bridge as the link to the next life. Less seriously we refer to past experience as 'water under the bridge'. You may delay by saying 'I will cross that bridge when I get to it.' Feeling a bit philosophical? Then life consists of 'many bridges to cross'. Options reduced? You have 'burned your bridges'. The imagery is reflected in popular music: Simon and Garfunkel sing, 'Like a bridge over troubled water', and Don Maclean wrote and sang a song called 'Don't burn the bridge'.

In Chapter 2 I quoted the Prime Minister of New South Wales, Australia, Jack Lang, when opening the Sydney Harbour Bridge, 'the bridge of understanding among the Australian people will yet be built.' We speak of building bridges of understanding between different nations, cultures, and religious faiths. In short the bridge as a link is almost as basic to the way we think as the ancient elements of earth, water, air, and fire.

People bridges depend on how we are as individuals and how we express ourselves. We do that through the actions of our 'body language' and the words of our prose, poetry, drama, art, and music. The 'chemistry' between people *emerges* from those expressions. Clearly prose, poetry, and drama are built on letters, words, and sentences of natural language. Physical bridges can be public art as we saw in Chapter 5—but art can also sustain people bridges as the artist seeks to communicate an idea through daubs of paint or the material elements of a sculpture. Music is built up in layers from notes, chords, bars, phrases, periods, sections (a musical idea—verse, chorus, refrain), and movements to make a composition.

Each of these kinds of expression use a hierarchy of symbols, materials, and sounds and each has an impact. The effect is not a direct property of the individual letters, words, notes, chords, or daubs of paint but *emerges* from the interactions and relationships between them. The interactions create the quality of content and performance of a people bridge.

The physiologist Denis Noble, in his book *The Music of Life*, uses music as a metaphor for the complexities of interactions at the various levels of an organism from genes to proteins, through cells, tissue, organs to the whole.[1] He describes listening to a CD recording of his favourite piece of music, a Schubert piano trio. The digital code on the disc is transferred by the laser reader and converted into a signal which passes through the amplifiers and the loudspeakers and turned into sound waves that travel across the room to his ears. He describes how he is so caught by the emotion of the music of the moment that he cries.

If we look at the processes that turn his CD into beautiful music we can see what seems to be an inevitable stream of causes and effects that we can trace directly from the digital code on the CD to his crying. So can we therefore infer that the code caused him to cry? No one would seriously argue that. It is obvious that without the genius of Schubert, the inspirational playing by the trio, and his emotional state at the time, there would be no crying.

Noble points out that it is just as big a mistake to think that the DNA in your genes causes you to be the person you are. He wants to discredit the popular view that once we know the genetic code then we'll know the person. As a systems biologist he sees an organism as different levels—but the levels interact in complex ways. He is arguing against the traditional scientific view called reductionism. This is the view that, in effect, says that if we know all about the DNA in the genes then, through an inevitable chain of cause and effect we will know all about the organism. It is as if the impact of a poem or a novel can be known just from the 26 letters of the words.

The book of a bridge is analogous to the book of a piece of art or an organism. You'll recall that the letters of the book of a bridge are the constituents of materials such as the sand, cement, and aggregate that makes concrete or the iron, carbon, and other elements that make steel. In an organism the letters are the complex molecules of DNA in genes, RNA, and many others such as water and lipids which combine to make the 200 or so types of cells of the human body.[2] We said that the words of a bridge are the materials such as steel, concrete, and timber; in an organism they are tissue such as flesh, muscle, bone, and blood. The sentences of a bridge are components such as masonry, beams, cables, and bars; in an organism they are organs like heart, lungs, liver, and brain. The paragraphs, sections, and chapters of a bridge are combinations of BATS; in an organism they are the complex combinations of organs that make up the muscular-skeletal system, the circulatory system, the respiratory system, the endocrine system, the immune system, and the reproductive system.

However, it is the complex interaction of the processes at the various levels that create the *emergent* characteristics. For example, Noble describes his own research into the mechanisms that cause the heart to beat. In one experiment the system is a muscle pacemaker cell with components which are protein molecules that channel the electrically charged ions to create the rhythmic activity. He describes how the components, the proteins, contribute to the behaviour of the system, the cell, but then the system feedback alters the behaviour of the components in a set of loopy processes.

Now you could be forgiven for thinking that physical bridges made of masonry, timber, steel, and concrete are much simpler than biological organisms and you would be right. Nevertheless the same principles apply. The reason we think bridges are simpler is because scientific reductionism has worked very well for them. Indeed it has worked so well that we know that the science is true. But do we?

There is a question and it is a very big one. Why does science seem to work sometimes and not others? Take the acid rain problem, for example. Acid rain corrodes bridges although the damage to trees and other plants is more serious. Rain is naturally slightly acidic because carbon dioxide dissolves in rain drops to give carbonic acid. If there are strong emissions of sulphur dioxide and nitrogen oxide from industry then the rain can become even more acidic. This became a really big issue in the 1980s. Acid rain doesn't recognize national boundaries so weather patterns and pollution from one country can harm trees and plants in another. For example, Canada thought the USA was polluting its forests and lakes. You may have seen photographs of dead trees killed by acid rain. The impact is dramatic. The arguments raged and eventually action was taken. Advanced filter technologies were developed and laws have been passed in Europe to force industry to use them. The problems have therefore been much reduced. However, in Asia this has still to be done. China in particular has a strongly expanding industry and relatively cheap and plentiful supplies of coal. If environmental standards are low, then acid rain is the result. Not only are trees and lakes polluted but people's health is affected and the infrastructure of buildings and bridges is damaged. Now there are signs that this is being recognized in China and changes will come. The story is one of hope.

When there is a gap between what we know (the chemical reaction) and what we do (pollute), then unintended consequences can occur (trees killed). If we understand the problem properly and we have the collective will to communicate between different professional disciplines, we can act and build bridges over the gaps and deal with the problem. Once we recognize that there is an issue we can take steps to deal with it—as long as we have time and the situation is indeed reversible. In the wobbly bridge and in acid rain the gap between what we know and what we do was and is being bridged—but not before some embarrassing and troublesome unintended consequences.

So one reason why sometimes science works and sometimes it doesn't is that our understanding is always incomplete.

It is undeniable that reductionism is powerful and important but it is also undeniable that it isn't sufficient for dealing with really complex systems. The many successes in bridge building are impressive but we actually understand less than we sometimes think we do. The scientific models are not absolutely true in all contexts; rather they are contingently true in specific contexts as we said in Chapter 6. We need new ways of thinking with science. We now realize that principles such as the safe theorem save us from embarrassment. The predicted stresses and strains in actual bridges are broadly as we expect but the complexities of real life mean that the relationship is approximate and uncertainty is significant. There are many phenomena that have only partly yielded to reductionism such as the fatigue behaviour of materials, the nonlinear dynamic vibrations of the many levels of a hanging bridge, and the detailed behaviour of engineering soils like clay. The principles that Noble applies to organisms also apply to bridges.

But for bridges there is more. All physical man-made systems are built to work in the human context that gives them meaning—but that context really is complex. And that's not all—we are embedded in the complex physical and natural world we call the environment.

In Chapter 5 we used the word 'hard' to mean something definable and measurable, and we used the word 'soft' to refer to something difficult to define. Traditional science is the science of hard systems based on reductionism. As Denis Noble says, many of the latest developments in the science of the human body have been gained through a reductionist hard science of biology and biochemistry. He argues that we now need to recognize that organisms are actually so complex that reductionism must be set in a wider systems perspective.

So from now on when I refer to a 'hard system' I mean a physical, material set of 'things'. When I talk of a 'soft system' I mean one involving human beings. For physical hard-systems bridges there is an action that creates a reaction which we understand through Newton's Laws. That understanding is sufficient to create bridges that work—they stand up. Soft systems are hard systems which contain multiple layers of human intentionality. At its simplest, intentionality is having a purpose, aim, or goal. It is this multiple layered interacting intentionality that makes soft systems so complex.

As Noble says, we must not abandon reductionism; rather we must see the emergent whole as well as the reductionist parts.

Reductionism on its own has been enormously successful for hard systems but it has reached its limit for some of our most difficult issues. It is unlikely ever to succeed for soft social systems.

Bridges are built for people. A simple example illustrates the complexity of soft systems. 'It must never happen again.' A phrase always genuinely meant—too often repeated. Media publicity of tragedies can be intense as people look for someone to blame. Whether bridge collapses, undetected child neglect, errors in medical treatment, computer records lost by government officials, or deaths from friendly fire in warfare, any loss of life is dreadful. Naturally we want it never to happen again.

Unfortunately, and all too often, history does repeat itself. Despite all kinds of enquiries, including formal commissions of inquiry, it sometimes seems that we want to learn the lessons but we just fail to implement them. Tragedies just keep on occurring.

Bridge failures are no exception. Thankfully they are very rare—but they do happen. We have looked at some in earlier chapters.

Millions of people believe in perfection—mainly through their religion. But most realize it is impossible to achieve in practice. It is something to aspire to but there is always a margin between what we do and absolute perfection. We always fall short. So what is realistic?

Bridge builders must be pragmatic—they must build bridges that will work. Pragmatic systems thinking is natural for bridge engineers but this is not the same as the American philosophy of pragmatism.[3] Bridge builders use practical rigour, as we discussed in Chapter 6, which is not the same as scientific or logical rigour. Logical rigour is necessary but not sufficient for practical success because practical knowledge is almost always incomplete.

Bridge engineers must be pragmatic about people bridges too as they form and reform teams for various projects. The hard practical fact is that no one can guarantee that disaster won't ever strike again. But while we may not be able to achieve perfection, we can make the risks acceptably low.

Yet the situation is perverse. When disasters are rare then decision makers tend to push the issues to the bottom of their priority list. Then there is a danger of neglect. That is just what has happened to the maintenance of bridges. Almost all countries in the world have a decaying stock as I mentioned in Chapter 4. Just to recap—the American Society of Civil Engineers says that in 2003 27% of the 590,750 bridges in the USA were structurally deficient or functionally obsolete.

The I-35 bridge referred to in Chapter 4 collapsed on 1 August 2007 in Minneapolis. But this bridge wasn't being neglected. It was inspected annually, indeed on the last occasion in May 2007.

There were lessons that had been learned previously. For example, the I-794 Hoan Bridge Milwaukee, Wisconsin, USA, was a tied arch bridge, built in 1970–2 and collapsed in December 2000. The bridge was being monitored when cracks had been discovered at a joint detail that had three-dimensional tension, i.e. tensile stresses in all three degree of freedom directions at the same time. The temperature was very low when it collapsed, tending to make the steel more brittle. The technical information about the incident was fed back by the US Federal Highways Administration to the Departments of Transport in the individual states. A two-day workshop was held in

September 2001 with more than 100 transportation officials from all over the USA.

Whilst the technical details at the I-35 were slightly different, the message from the Hoan Bridge disaster was clear—bridges susceptible to fatigue damage need to be carefully monitored. The I-35 Bridge was being monitored but it still collapsed—why?

Before we attempt to answer that, let's look at the UK situation. The UK has around 160,000 major highway bridges, many of which were built in the 1960s and 1970s or before.[4] Most were built in reinforced or prestressed concrete. Weathering and de-icing road salts have corroded the steel which expands and the concrete spalls. In November 1987 the UK government started a 15-year programme to assess and strengthen bridges.

One of the problems was, and still is, dealing with bridges found to be substandard. In August 2002 an important review of the whole programme found that there were an appreciable number of substandard structures for which there was no evidence of them being managed properly. The report concluded that bridge owners, 'where they have sub standard structures ... are not always managing the risk of collapse in a safe and auditable manner'.

The problem is not restricted to USA and UK. Around the world there is an enormous stock of bridges to be maintained.

Then there are the surprises. As I have been writing this book the fate of the Glasgow Clyde Arch Bridge featured in Chapter 1 has dramatically changed. After initially being feted as an innovative and successful bridge, in January 2008 one of the cable connections snapped totally unexpectedly. An immediate inspection revealed a second faulty connection. In March 2008 five bespoke saddle frames were installed to allow all of the connections to be replaced. The bridge was reopened in June

251

2008. At the time of writing poor manufacturing of connection holes and faulty steel are being blamed for the failures.

The I-35 was another surprise. The US National Transportation Safety Board (NTSB), in November 2008, published the report of their investigations. They found that regular inspections had been diligently carried out. Every element of the bridge had been photographed and checked for fatigue cracking. The bridge was being closely monitored for fatigue. So this was not the cause of the collapse as many had first thought. It emerged during the inquiry that some bowing of the gusset plates used to connect the members of the truss was noticed by the inspectors. However, they did not mention this in their formal reports. They reasoned that the bowing had been there since the bridge was built because they knew (or thought they knew) that gusset plates were generally over-designed. In any case, design and construction issues were not part of a maintenance inspection.

The NTSB inquiry team concluded that the bridge collapsed because particular gusset plates had failed—they had not been properly designed. Also the weight of the bridge had been increased through previous modifications and on the day of the collapse there were extra concentrated traffic and construction loads. The team concluded that the bridge designer's quality control procedures failed to ensure that appropriate calculations were performed and the design review was inadequate. The generally accepted practice of the time gave inadequate attention to the gusset plates.

So the warnings about possible fatigue damage were not missed. Indeed they were the focus of attention. Unfortunately, this meant that what turned out to be the actual cause was missed.

The story is further evidence that the prevailing culture of engineers and engineering is a major factor. Engineers of all kinds are educated and trained to focus only on the hard physical material nature of systems. Their education is based on an unstated assumption of scientific reductionism, which as we have already recognized is insufficient. It is therefore a shock to many engineers, after graduation and into their first job, to realize that much of their day-to-day professional work is actually spent on dealing with messy and uncertain problems—and they are surprised by how much time they must spend dealing with people bridges. They must learn quickly. Quite naturally they see people bridges from their technical perspective. Consequentially their understanding of people bridges and soft systems is sometimes not as well developed as it might be. They may not appreciate the difficulties of delivering joined-up information sources in quality control. They may not appreciate that critical problems may not be the ones you are most concerned about. Barry Turner, the sociologist whose ideas about 'incubating failure' we examined in Chapter 6, called this the 'decoy phenomenon'. There may be a tendency not to give sufficient credence to warning signs because of the natural human tendency to think that the worst cannot happen.

The root of the issue is that there is a gap—indeed a gulf—between the way most of us think about the hard and the soft. It is a gap that needs to be bridged. We need to find ways of 'joining them up'. We need to think of the physical and the human as one integrated whole.

In case you think this is only a problem for bridge builders let's just look at one different but related example—the idea of 'joined-up government'.[5] Tony Blair's first 'New Labour' UK government wanted to get departments and agencies to work more closely together. They saw a need for better collaboration across organizational boundaries to deal with shared issues. They wanted to improve the flow of information to deliver better services with a focus on the needs and convenience of the customer rather than the provider.

So the government introduced major programmes to attempt to modernize many parts of government bureaucracy including the national health service and the criminal justice system. The commitment was set out in the 'Modernising Government' White Paper (1999). The then Minister for the Cabinet Office, Jack Cunningham, wrote in the introduction to the white paper: 'We need joined-up government. We need integrated government.' One of the three stated aims of the new policy was to ensure 'that policy making is more joined up and strategic'.

It was an ambitious project. They were setting out to reform the very processes by which government itself works—especially cross-cutting issues that all organizations find difficult to handle.

So what happened? The 'joined-up' phrase was used for some time but then got quietly dropped. Perhaps that is understandable since the issues are tough—but they haven't gone away.

In February 2006 Simon Caulkin wrote in *The Observer* newspaper,

> New Labour was right in 1997 that joining up public-sector management was its biggest challenge. Unfortunately, in 2006, it still is ... But if concentrating on parts of the problem initially seems easier, it always ends in tears, adversely affecting the system as a whole, raising costs and making it harder and harder for managers to see the link between causes and effects.[6]

He continued:

> everything is part of a bigger system so it is difficult to know at
> what level to tackle an issue and how best to measure its per-
> formance ... the NHS, for example, can't simultaneously meet all
> its performance targets despite absorbing record amounts of
> money because without a systems view of improvement, per-
> formance goals can be met only at the expense of missing others.[7]

Performance targets have received a bad press. Charles Goodhart, once a chief adviser to the Bank of England, observed that a
measure stops being useful as a measure when it is used as a
target. He said that it can be one or the other, but not both.
So when government tries to regulate a group of financial
assets then the measures they use can no longer be relied on as
economic indicators because institutions just invent new cat-
egories of assets. In other words, like Heisenberg's uncertainty
principle of quantum mechanics, Goodhart's observation says
that measuring a system usually disturbs it.

However, his argument misses something essential. He was
referring to financial institutions, government, and other organizations that do not share a common purpose—indeed they have
different purposes. Goodhart's observation fails if all players in a
process are working to the same end—aiming at the same targets
just like footballers wanting to score goals—a prime example of
common purpose. No one could seriously argue that scoring
goals cannot both be a measure of success and a target.

Targets are central to what Professor Michael E. McIntyre
of the University of Cambridge has called our 'audit culture'.
He comments that this seems to have been driven by three
very reasonable principles of fairness, objectivity, and prudence.

Fairness says that everyone must be treated in the same way. Objectivity requires that all assessments must be based on numerical measures. Prudence seems to suggest that no one can be trusted so we need to audit them. But who audits the auditors? How can the unmeasurables such as professional ideals and ethics, ambition, curiosity, enthusiasm, room for creativity, and willingness to share be rewarded?

He concludes,

> There is actually no alternative to reliance on trust—and to rebuilding trust where necessary. An advanced society will recognize this explicitly and live with the risks. It will use auditing resources in new and cost-effective ways by concentrating them not on trying to monitor and measure everything but, rather, on checks and balances against gross human failings.[8]

One of the UK government's more successful initiatives has been that of 'Rethinking Construction'—now rebranded as 'Constructing Excellence'.[9] While at its best the UK construction industry is excellent and matches any in the world, too often it has underachieved with low profitability, too little investment in capital, research and development, and training. The industry is highly fragmented and too many clients have been dissatisfied with performance. The Deputy Prime Minister John Prescott invited a group of clients chaired by Sir John Egan to suggest improvements. They reported in 1998. Since then a number of changes have been made and achievements monitored. The five key drivers for improvement were committed leadership, focus on the customer, integrated processes and teams, a quality-driven agenda, and commitment to people: in other words—joined-up construction. A joined-up systems thinking approach to these issues is being promoted and used in UK and other parts of the world such as Hong Kong and New Zealand.[10]

Joined-up thinking is difficult and controversial. Yet the best bridge builders must do it—if they don't, their projects will fail. Joined-up thinking is about cooperation, coordination, communication, integration, and synergy.

So let's approach this by recognizing that building bridges is, at its core, about practical problem solving. By problem solving I don't mean puzzle solving like a crossword or Sudoku but the creative challenge of identifying issues and meeting them to deliver a significant human need. At the same time we should recognize that the flipside of a problem is an opportunity—so bridge building is also about finding and taking opportunities.

Let's start by looking at a simple problem first suggested by Peter Senge in his wonderful book on systems thinking, *The Fifth Discipline*.[11] What functions do you actually perform when you turn on a tap to fill a glass of water?

Before reading on please pause a moment. Think about all of the important steps in the process of turning on the tap and filling the glass. Now, if you can, write them down and do this before reading on.

Perhaps you wrote a list or drew a diagram similar to the one in Figure 48a. If so, you are not unusual because most of us are trained to think in straight lines. But if you reflect for a moment, turning on a tap is not a linear process it's a cyclical one. As you turn the tap you are constantly watching the water level and adjusting the water flow until the level gets to what you want. So the stages are redrawn as processes in Figure 48b, where the arrows show how one process of change influences another. The result is Figure 48c.

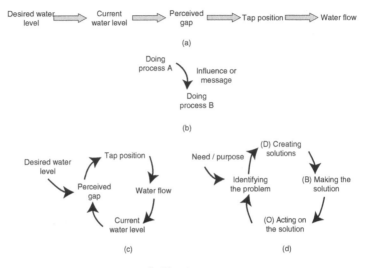

FIG 48. Turning on a tap

We start with a need—the desired water level. We perceive a gap between the water level and the one we want so we open the tap by adjusting its position. As a consequence water flows and the water level changes. We see a new gap and adjust the tap position again. We go round and round this loop until we get the level we want and then we turn the tap off. Simple! Yes but it's more complex than perhaps it first appears—as anyone who has tried to design a robot to do this kind of task will tell you.

In general, solving a problem is cyclical and the stages are subprocesses, as shown in Figure 48d. The present participle, 'solv-ing', has been used to make this clearer. Each of the five stages needs to be managed to success if we are to manage the original problem to success. The subprocesses are understanding the need, identifying the problem, creating solutions, making the solution, and acting on the solution. As we go around this cycle time moves on and the situation changes. The result is that when we reach the end of the cycle we now have a new state of affairs and new problems. So we must start the cycle yet again. The cycle is not just a circle, but a continuous spiral through time.

But there's more. Each subprocess will have sub-subprocesses—indeed they are all parts and wholes—holons (Chapter 4). So there are process holons at every level of definition all going through spirals. From the top process holon right down to every detailed process holon there are continuous interdependent problem-solving spirals.

So now let's look at these five stages in a bit more detail. The first is the driver of the spiral cycle—need or purpose. One way of finding this is to simply ask 'Why?' Why are we doing this? Why do we pour water into our glass? We need a drink. Why are we building a bridge? We need to cross the river. Need and purpose are the voltage that drives the current of change around the cycle, or it is the gravity force that makes the glass you are holding fall to the floor if you let go.

The second stage follows directly on—identifying the nature of your problem. What level of water do you need in the glass? Perhaps a ferry is all that is needed to cross a river? Do you need a footbridge or a highway bridge? Just what are you trying to achieve in solving this problem? What are your aims and objectives?

The third stage is deciding what to do about your problem, i.e. how far to turn the water tap. In general it is about creating possible solutions and deciding on the criteria for choosing one of them. We will call this (D) for *designing* since for a bridge we must decide on the form, materials, and detailed structure.

Designing is an 'opening-out' creative process where ideas are suggested and developed. Note that designing is more than simply creating something that looks good. It is also about choosing the criteria of what is good or deciding why one solution is better than another. Those criteria must reflect a whole range of needs. So designing is an evaluative process of deciding.

The fourth stage is doing something or something happening like the flow of water through the tap. The 'state of affairs' changes. We will call this (B) for *building*—making the solution we have designed happen. We actually change our reality—we implement our solution: we build the bridge.

The fifth stage is acting on the solution—we will call this (O) for *operating* or *using*. We work with the changes we have made, we operate our solution. We monitor the changing water level. We use the bridge we have built. Naturally doing this leads to a whole new set of problems. For example, we must continually maintain the bridge and, at the end of its life, we must demolish it.

So now let's consider the building of a bridge in more detail. We have three interdependent needs or requirements to satisfy—firm foundations, strong structure, and effective working. Success in all of them together creates a successful bridge. In other words, the three process holons are together necessary and sufficient for the success of the bridge. But because the three processes are interdependent they need integrating. So our top-level need and purpose is to make sure that they are integrated—it's a kind of meta-purpose, i.e. one which surrounds the more obvious need and purpose of building a bridge to bridge a gap. So at this top level the need/purpose and the process of identi-

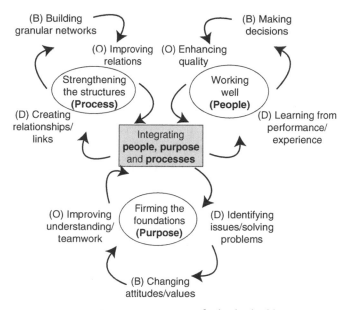

FIG 49. Integrating processes for bridge building

fying the problem in Figure 48d have become the single integrating process. That is why it is at the heart of the three interacting problem-solving cycles in Figure 49.

Let's now look at each of these requirement cycles in turn.

In the preceding chapters we have examined the way foundations are made for physical bridges. We have recognized that foundations hold the rest of the bridge in place. But how important are the foundations for a good people bridge?

One word captures the idea of a good foundation—purpose. For people bridges, a good foundation is a firm foundation for everything else in life. It holds the rest of our lives in place—it underpins all that we do—it gives us a sense of place and purpose. It is in very large part provided by our education and upbringing. It is about laying down the basis of the way we live with others to provide a quality of life. This hopefully includes, for most of us, religious and cultural tolerance and the many other things that lead, ultimately, to our own self-fulfilment and happiness.

In this wider sense, foundations for successful physical bridges are not just the groundworks that we have discussed in earlier chapters. We must include the ideas that enable us to create and appreciate objects of beauty which work in harmony with their surroundings. Bridges can create a sense of awe and, as works of art, a sense of the spiritual. Bridge building can help us appreciate holistic quality—which is not just about aesthetic form or about functional efficiency—but a whole raft of requirements including sustainability.

The second requirement for a good bridge is a strong structure. One word captures this idea—process. In a physical bridge the process is the flow of internal forces through the physical set of interconnected pieces of material which is the structure. The structure of our own people bridges consists of our relationships and links with others. In other words, a structure is the form of a process. Without structure nothing can work—everything must have structure—the only question is whether it is a good structure, i.e. fit for purpose.

Building a bridge is a team effort so the structure of the people bridges required to build a physical bridge is the structure of the team. Architects, structural engineers, geotechnical engineers, project managers, welders, steel fixers, and many more professions and trades are involved. The project leaders must build a great team in order to build a great bridge.

Just as we can span a physical divide, such as a river, with a bridge of steel, so we can span deep intellectual, emotional, religious, and cultural divides, such as self-interest, fear, and intolerance, with bridges of understanding. That is not to suggest that bridges of understanding are easy to create. Tragic human stories, including child murders such as that of Victoria Climbié,[12] often involve the effects of a remarkable lack of joining-up.

The third requirement is that bridges must work well. One word captures this idea and that is people. This cycle represents the performance of the physical bridge (which, although it is a physical object, is perceived, understood, and responded to by people) and the performance of people and communities through living their daily lives. Ultimately all bridges are there to work for people.

Builders of physical bridges depend on learning from practical experience—from finding out what works and what doesn't. They look for evidence that they can depend on. They realize that they have a big responsibility for public safety—if a bridge fails, then people will probably be killed. Bridge builders tend to have a well-developed ethic that values the honesty forced on them by Mother Nature because you just cannot fool her. Gravity will do what it does no matter what spin you put on things. A very large part of the skill of a bridge builder comes from 'hard-fought' experience.

So what is the nature of the process needed to integrate or join up these three processes of firming the foundations, strengthening the structure, and working well? How will we know if we are successful?

We first need to understand what we are aiming at. I define a joined-up system as one where we get the right *information* to the right *people* at the right *time* for the right *purpose*, in the right *form* and in the right *way*.

We can unpick this rather long definition using *what, who, when, why, where,* and *how.* Think of the information as answers to questions starting with the word *what*—simply, what is the information we need and may want to transmit? Think of the people as the answers to the questions *who*—simply, who is involved? Think of the time as answers to the questions *when*—simply, when is it needed? Think of the purpose as answers to the questions *why*—simply, why is this needed? Think of form as answers to the questions *where*—simply, what form should the information be expressed in and what assumptions should be made about its context? Finally, think of the way as answers to the questions *how*—simply, how should the information be transformed? So the definition just boils down to asking six questions: *what, who, when, why, where,* and *how.*

This is all very easy to say, but not so easy to deliver, especially across all processes in the entire system. If there is a single deficiency in any of these requirements then there will be a lack of joining up. All it takes is a message, or piece of information, that doesn't get sent or received, or is poorly formulated, incomplete, misleading, or without adequate justification.

There is another way to see this. Joining up is not just about the flow of information between people; it is also about the flow of a physical material. Think of the flow of water around your house in the network of pipes or from the waterworks along a network of pipes to your tap. Think of the electricity flowing around the ring main of your house or on the national grid. Think of the flow of forces through a bridge structure. Any lack of joining up in these systems can mean no water at the tap, no electricity at the wall socket, or a bridge falling down because the forces become unstable. So we are able to create successful joined-up systems—but they are ones that have yielded to a scientific reductionist approach.

At first sight these two interpretations of joining up are very different—one about people and one about physical things. But it is imperative that we link them. That way we can see better how people are connected to their environment. We need to do it to minimize the unintended consequences so often at the root of why things go wrong.

You may recall that in Chapter 2 I introduced you to Karl Terzaghi, the father of foundation and geotechnical engineering. At Harvard University Terzaghi met a young man called Ralph Peck. Together they developed a joined-up way of working called the 'Observational Method'.

Their first meeting in 1938 developed into a life-long relationship. Peck recalled later that he was asked to help Dr Terzaghi with the correct English words for some statistical terms. Terzaghi simply wrote down the formulas for such a thing as the standard deviation and asked what we called it in English. Peck said, 'I left feeling quite exhilarated, because I had talked to the great Dr Terzaghi.'[13]

Soon after, Terzaghi was asked to go to Chicago and speak to the local section of the American Society of Civil Engineers. He chose to speak on 'The Danger of Tunnelling beneath Large Cities Founded on Soft Clays'. He evidently scared the audience because he was immediately asked to work for the City. Nothing happened for about a week but then he got a telegram saying, 'Your terms are accepted, please send your man.' But Terzaghi didn't have a man! He didn't even know any people to recommend. After consulting a few people he asked Ralph Peck.

Ralph Peck was born in Winnipeg in 1912 but he was never a Canadian. He worked on the Bronx Whitestone and the Henry Hudson Bridges in New York City. Unfortunately, the recession of 1937 came along and the bridge company ran out of work. He was suddenly unemployed. So he decided to try his hand at the new subject of soil mechanics. Professor Casagrande offered him work at Harvard and there he met Terzaghi. He never looked back. He went on to spend three decades at the University of Illinois pioneering new work in foundations, ore storage facilities, tunnel projects, dams, and dikes.

Construction had already started on one of the station sections when the young Peck arrived in Chicago in 1939. Large settlements of the buildings and of the streets over the advancing tunnels were already occurring. The contractors didn't want to change what they were doing—they insisted that the settlements were purely coincidental and had nothing to do with them. Terzaghi and Peck began to measure the movements of the ground around the tunnels. They soon established that the volume of the settlements of the street was roughly equal to that of the movements measured inside the tunnel. The contractor's case was destroyed.

Terzaghi and Peck knew that the first key to any chance of a joined-up approach between the parties to the contract was to establish agreement about what was happening. In other words, everything important that could be measured had to be measured. Terzaghi specified the measurement he needed and Peck wrote to him every afternoon, detailing the results and how the job was progressing. Terzaghi examined each report and kept close track of the job with numerous visits. Peck said, 'Working with Terzaghi was stimulating, demanding, and exhausting. He worked almost all day and night and he expected everyone else to do the same.'[14] At the same time he was kindly and understanding—everything seemed like a new adventure. Soil mechanics was a new subject and almost any observation of behaviour in the field provided previously unknown information.

While all this was going on, Terzaghi was writing a book in which he set out a new theory about the earth pressures against vertical circular shafts. Unfortunately, results from Chicago did not remotely agree with the new theory—so he scrapped it.

Then Terzaghi and Peck began to prepare a book together—it took seven years to write. 'On the half dozen projects in which we cooperated after the Chicago Subway the arrangements were always similar,' said Peck.

> It was my job to organize the exploration, to set up and supervise the field observations, and to describe what was happening in connection with the construction or behaviour of the facility. Terzaghi analyzed the results on a continuing basis, made recommendations to the client, and made further suggestions for observations that should be carried out.[15]

Terzaghi and Peck's Observational Method, which developed out of this collaboration over a number of projects was a technique well before its time. It has been used by many geotechnical engineers—but it has not received the acclaim it deserves. Put simply, the idea is to monitor the construction as it proceeds and then to act according to what happens—pretty much common sense you might say. In a large construction project some difficult decisions must be made in advance even when there are large uncertainties about the nature of the ground. The Observational Method requires a different approach to the one normally used. The first design work is done with the limited information available. At the same time, contingency plans are drawn up as to what should happen if things do not turn out as expected. Specific performance parameters are defined and carefully monitored. For example, the settlement of a section of ground or the movement of a wall may be measured at regular intervals. If the values exceed a stated threshold level then specific actions are triggered. This may all sound quite straightforward but it is tricky to achieve in practice.

Peck later commented, 'The Observational Method, surely one of the most powerful weapons in our arsenal, is becoming discredited by misuse.' He became very concerned that too often it was invoked by name but not in deed. He said that simply deciding what to do and then observing the consequences is not the Observational Method. Thorough investigations must be carried out to establish what to do if unfavourable conditions are actually found.

His remarks were foreboding. In 1994 a tunnel, being dug for the Heathrow Express Rail Link, collapsed. Peck's very warnings had not been heeded. The Heathrow contractors were relying on the Observational Method without understanding the implications of not following it through thoroughly. As Peck stressed, it is imperative that the engineers must devise in advance a way to deal with any problem that may be revealed by the observations. Then when the observations are taken, the results must be acted on. At Heathrow the results were not examined properly and no action was taken when required. The consequences were inevitable.

The engineers at Heathrow knew that the world constantly changes. What they didn't realize is the importance of careful monitoring of the factors critical to success.

Change is flow—continuous movement. Managing change is an issue in so many aspects of life. When we think about change, no matter the timescale, we try to explain and understand it. Inevitably we look for a reason or a cause. The idea of cause and effect is ingrained in our practical everyday thinking but has been a concept that has troubled philosophers ever since Aristotle. Common sense tells us that if a wine glass falls to the floor (effect) it is because someone or something dropped or pushed it (cause). But the effect may then become a cause of a different effect (wine staining the carpet). So causes and effects are not simply distinguishable. Likewise common sense doesn't help when we postulate that we humans exist (effect) because God made us (cause). Such statements are a matter of faith.

The world is a complex place and there are limits to what we know and what we can know, as we recognized in Chapter 4. In science we routinely ascribe causes to effects. But even in science there are limits. Although it is clear that water flows downhill (effect) because of a difference in pressure (cause), and voltage (cause) is the potential that drives the flow of current (effect), there are many situations that are not so clear. For example, in quantum mechanics many physicists would argue that a subatomic particle is never at rest and has fluctuations that have no cause.

In everyday life and in the practical world of bridge building there are many examples where causality is not simple. Did the young engineer who made the calculation error in the story of the failure of the Second Narrows Bridge in Chapter 3 cause the collapse? One short answer is yes, because it was his mistake that led to the decision not to stiffen the grillage beam which failed. Another short answer is no, because his mistake should have been discovered by his boss. Everyone makes mistakes at some time; that is why we work in teams and why we introduce checks and balances. All of the bridge collapses described in this book are complex with no simple cause and effect. Nevertheless the stories are narratives of change. As we saw in Chapter 6, sociologist Barry Turner said that many man-made disasters don't just happen, they 'incubate'. Accidents and disasters are usually, but not always, complex processes of change.

So how can we deal with all of this complexity in a joined-up way? One way is by changing the way we think—to become systems thinkers. We have already met the three essentials for a joined-up systems thinking approach, *thinking in layers, making connections,* and *identifying processes.* So these could become regular habits of our thinking.

The first habit to develop is to think about everything as having layers. This way you begin to see parts and wholes, holons (see Chapter 4), in levels of understanding. From letters to books, from DNA to whole organisms, from individuals to nations, from subatomic particles to the cosmology the systems thinker does not see any level as being more 'fundamental' than another. Scientific reductionism, in this respect, is the misleading idea that the deeper and more detailed you examine a system the more fundamental you get and the more likely you can build the whole behaviour of a system from its detailed parts. It is the view that we can know the man just from his genes. It is the view that we can have a 'theory of everything', i.e. know the universe from a unified set of forces.[16] A systems view is quite different. It says that you choose to understand and deal with the level appropriate for your need. Then you move around the levels as required.

The second habit to develop is to look for and to make connections. Systems thinkers understand things by seeing interacting connected systems of holons creating emergent properties at the next layer above. Our abilities to walk and talk are perhaps the best examples—none of our subsystems can walk and talk on their own but working together they do. The highly connected neural connections in the brain create emergent consciousness. Well-maintained connections between people make for good people bridges. Well-maintained connections between physical elements make for good physical bridges. Well-maintained connections between people and physical bridges make for good systems.

The third habit is to always look for the processes. Processes define what people do and how physical objects behave. They define structure. As we saw in Chapter 5, to do this you really need to throw away old ideas about the nature of what a process is and adopt a completely new view, one in which you start your thinking by seeing everything, including objects, as a process. You do this because, at its simplest, everything changes through time.

If we put these three habits together then we see a system as consisting of interconnected process holons. They are the structure for cyclical interactions, and through those interactions we get emergent properties.

Even if we accept that these habits are useful, how do we set about using them? For example, if we are looking for processes, how do we recognize them? How do we categorize them?

One way is through the grammar and letters to the chapters of the book of physical bridges. Denis Noble, as we saw earlier, has used the same idea in his biological music of life.

But how can we do it for soft systems? As we said earlier a soft system has multiple layers of intentionality. Intentionality creates purpose—so our starting point is always to be clear about our top-level purpose.

There is no single best way to classify processes. Earlier, integration was our clear purpose. Whichever way we choose must be practical and useful. One such way is to use the acronym BCIOD + R, which contains I for integration. This is a system developed by Patrick Godfrey and others when planning the building of Terminal 5 at Heathrow Airport. Let's see how it works on our own day-to-day living as well as on how we build a bridge.

The first category, B, is for business processes, which at a personal level are mainly financial processes. In bridge building B is for all aspects of the financial resource management. Next comes C, which stands for customer processes, which is about the people you relate to, your employer or anyone you undertake to do something for or make promises to. In bridge building C stands for all stakeholders including the client and the general public. It also includes the physical and natural world which will be affected by the project.

I is for integrating processes. These are things we do to 'oil the wheels', such as keeping a diary, maintaining our relationships and our network of friends. In bridge building I is for team building, building commitment and championing the cause, establishing consensus, and building trust by being open.

O is for operating processes, which are the way we run our day-to-day lives. In bridge building O includes all the processes that create the actual bridge—the things we must do to design, build, and operate it—then finally demolish it.

D is for delivery. These are the processes that deliver the things for which we are responsible and accountable to others, such as doing things for others, keeping promises, and delivering goods and services. In bridge building it is about getting the human resources with the required skills and competencies and the right kind of infrastructure such as IT networks and site provisioning.

We must recognize that any classification depends on context. One man's D is another man's O. In other words, a company will have operational O processes, for example, to manufacture bridge bearings, which will be seen as delivery D processes for the bridge itself.

Finally, R is for regulatory processes such as observing the rules of law, like driving on the correct side of the road. In bridge building it is appropriate contracts between the parties, clear policies on health and safety, and ensuring that the law is not broken in any way.

The stories of bridge failures have shown us that it's the integration of people, purpose, and process which is so often sadly lacking. Examples range from the enquiry into the whole construction industry mentioned previously,[17] to the report into the London bombings on 7 July 2005.[18] To see how integration is at the heart of success let's return to the three processes for building a bridge: firming the foundations, strengthening the structures, and working well. So let's look again at Figure 49 with its three interacting cycles.

Figure 49 shows that when basics are well thought through, awareness of the issues is high, structures are well designed and built and used well; then we can create an upward spiral of improvement. When things go wrong the lessons learned are fed back to each of the interconnected spirals and improvements follow.

In people bridges it is important to recognize that integration does not mean assimilation. Integrating is about bringing people into an equal association, working together for mutual benefit. It involves trust and respect—it is not about making everyone the same but it is about having a set of core values which we all live by. Assimilation is where minority cultures adopt the customs and attitudes of the majority. This is quite different and separate from the notion of integration.

What happens in Figure 49 when things go wrong? The upward spiral turns into a downward spiral of decline, damage, and loss. *Poor integration* at the centre of the diagram is sufficient but not necessary to make the three cycles turn into 'Confusing the basics', 'Inadequate structures', and 'Not working well'. Again the lower cycle underpins it all. Poor integration will aggravate confused basic values and partial thinking that reinforces the shared assumptions, attitudes, concepts, values, and practices of the group. Awareness of the issues and the value of others outside the group will be incomplete. This feeds back to make the poor integration even worse.

Poor integration results in structures vulnerable and susceptible to failure. With confused basics and inadequate structures, the operational process does not perform well. People attempt to respond and improve but quality is poor. Lessons to be learned are only partially heeded and not fed back and integrated into the rest of the thinking.

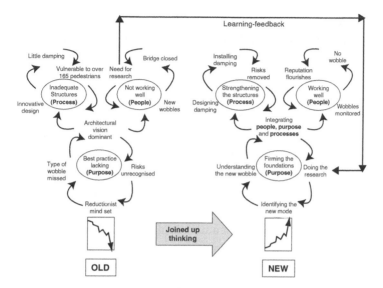

FIG 50. Feedback for learning

As a result the whole system is a complex of three interacting spirals of decline. It is unsurprising that it is difficult to untangle and turn around.

In Figure 50 I have attempted to capture the story of the wobbly bridge in Chapter 1. The left-hand cycles represent the original story. The arrow link between the 'Need for research' and 'Doing the research' is the learning—feedback. It is a step change in the relationships between the processes (from 'Old' on the left to 'New' on the right) forced by the closure of the bridge. The right-hand cycles are the new situation after the research into synchronous lateral excitation.

The difference between the two cycles, 'Old' and 'New', is new understanding. The 'Old' lower 'Firming the foundations' cycle is that best practice was lacking. As a result the risks of the new wobbles went unrecognized. The small amount of literature on the topic was missed and so the possibility of the wobble was not recognized. Without that recognition the process of actually designing and building the innovative bridge inevitably resulted in too little damping. Hence the bridge was vulnerable.

But now we know that many pedestrian bridges of very many different forms are also vulnerable. The difference is that they have never had large enough numbers of pedestrians on them to cause a problem. The Millennium Bridge was unfortunate in this respect since it had a very high pedestrian loading at its opening. Consequently the bridge wobbled and was closed down. The need for research meant a move back to the basics in the right-hand diagram. So the research needed to firm the foundations of our understanding of why the bridge wobbled had to be done quickly. As we know, synchronous lateral excitation turned out to have been exhibited in other kinds of bridges quite different in form from that of the London Millennium Bridge. The phenomenon was poorly understood and, despite the research, there is still much to learn. The way pedestrians respond to movements of the bridge is much more complex than simply walking in step (synchronicity) and is still not totally understood. New design rules are being written and a lot of research is still being done. However, once the new type of wobble was identified for the Millennium Bridge and the need for more damping was understood, then the new dampers were designed and installed. The result is a bridge that works well, and is much appreciated by almost everyone who sees it. Indeed it is a source of delight.

There is a general lesson here. Bridge building can help us understand better what we have to do to address the greatest issue that we humans have ever faced—climate change. Our ability to predict the future is limited. That is why there has been so much controversy surrounding the reports of the Intergovernmental Panel on Climate Change and what needs to be done.[19] One consequence is that many people shrug their shoulders and say that is up to others to do what's necessary. It is easy to think that your own contribution is so small as to be insignificant in the totality of this enormous issue.

Clearly on some issues it is the specialists who must act. For example, it is prudent to assume that physical bridges, like the rest of the built environment, will have to cope with more extreme weather events. Bridge builders have a responsibility to cope by developing even better methods of risk management. Likewise our political and business leaders would be prudent to assume that more turbulent times lie ahead. Again this calls for better long-term risk management.

If climate change is as urgent as the specialists tell us, then we do not have much time. A hard-systems analogy may help. In an electrical circuit, resistance is the dissipation of energy. War and internal conflict are all dissipating resources that ought to be invested in tacking climate change. So conflict is a resistance in the process of tackling climate change. Conflict reduces our capacity to act together.

It is clear that if we are to act collectively on the timescale needed, we will need new cooperation through new people bridges. But at all levels, from the letters to the book of the whole world, it's the people bridges that are so crucial and the hardest to build. Whatever we do, at whatever level we operate, unintended consequences will occur. Our best strategy is to absorb current reductionist thinking into a systems thinking evolutionary observational approach. As we work at each level at all of the issues we face we must be mindful of our effects at other levels. What each of us does might seem small in comparison to the whole but the emergent properties of what we all do together are what we should be focusing on.

Dealing with climate change urgently requires a shared clarity of vision and purpose at all levels with a sense of total integration. It requires us to expect the unexpected and therefore to be prepared to change and adapt.

Terzaghi and Peck never anticipated that their observational approach would be used outside the engineering of foundations. But their ideas are simple common sense. Their very practical method can be consciously applied to the world as a total system in a way that everyone can understand. In doing this we must monitor, measure, and look for evidence of important trends. Scientists are already doing this, e.g. measuring rates of melting of the polar ice, but we must do more—much more. We must do it for all processes at all levels.

Then we must integrate and interpret what is happening, which is far from trivial. The media have an important ethical role here within their commercial purpose. They must help people to understand statistical evidence rather than, as is all too often the case, to mislead with implications drawn from anecdotal stories. All of us then must plan what to do at many and various levels, for all foreseeable outcomes. But we must do that whilst expecting some complete surprises. As we do in everyday life we deal with surprises as best we can. The message is one of hope as several international consultants are embracing systems thinking as they face the challenges of sustainable construction.

The slogan 'Save the Planet' is the most misleading ever. The planet will survive—the question is whether the human race will. Bridge building has much to offer.

Endnotes

Chapter 1: Bridges are BATS

1. Andrew Walker, 'Lord Foster: Stormin' Norman', *BBC News*, Friday, 1 February 2002 [online]. Available at: http://news.bbc.co.uk/1/hi/in_depth/uk/2000/newsmakers/1796173.stm

2. Jonathan Duffy, 'The Trouble with Modern Architecture', *BBC News*, Monday, 12 June 2000 [online]. Available at: http://news.bbc.co.uk/1/hi/uk/787507.stm

3. Yozo Fujino, Benito M Pacheco, Shun-ichi Nakamura, and Pennung Warnitchai, 'Synchronization of Human Walking Observed during Lateral Vibration of a Congested Pedestrian Bridge', *Earthquake Engineering and Structural Dynamics* 22 (1993), 741–58.

4. A newton is the force that produces an acceleration of 1 metre per second per second on a mass of 1 kilogram.

5. The relationship between the stiffness of the bar and the elastic modulus of the material is as follows. The stiffness of the bar K is the force P divided by the stretch x, i.e. $K = P/x$. The force P is the stress s multiplied by the area A and the stretch x is the strain e multiplied by the length l. Thus $K = (s \times A)/(e \times l)$. Since the elastic modulus of a material E is the stress divided by the strain (s/e) then $K = (E \times A)/l$. In other words, the stiffness of the bar K is the elastic modulus E multiplied by the area of the bar A, divided by the length l, which results in the stiffness being

$$E \times A/l = (10 \text{ kN/mm}^2) \times (150 \text{ mm}^2)/(1{,}000 \text{ mm}) = 1.5 \text{ kN/mm}$$

as stated in the text.

6. In Chapter 6 we see that this equilibrium may be static or dynamic.

7. Then, of course, the whole bridge is an object embedded in and connected to its physical and environmental context and more than that it is embedded in its human, social, political, and cultural context.

8. The mathematics of coordinate geometry usually refers to x, y, z coordinates.

9. Ross King, *Michelangelo and the Pope's Ceiling* (London: Pimlico, 2006).

10. John Beldon Scott, 'The Art of the Painter's Scaffold, Pietro da Cortona in the Barberini Salone', *Burlington Magazine* 135/1082 (May 1993), 327–37.

11. Quoted in Jessica Rowson, 'New Bridge Competition Row', *New Civil Engineer* 6 June 2008. Available at: http://www.nce.co.uk/new-bridge-competition-row/1544039.article

Chapter 2: Underneath the Arches

1. You can see this painting at: http://en.wikipedia.org/wiki/Basilica_of_San_Francesco_d'Assisi

2. You can see this painting at: http://commons.wikimedia.org/wiki/File:Giotto_-_Legend_of_St_Francis_-_-06-_-_Dream_of_Innocent_ III.jpg

3. You can see this painting at: http://en.wikipedia.org/wiki/Image: Fra_Angelico_043.jpg

4. F. W. Robins, *The Story of the Bridge* (Birmingham: Cornish Bros, 1948).

5. Ibid.

6. 'Ponte Vecchio', *Wikipedia*. Available at: http://en.wikipedia.org/wiki/Ponte_Vecchio. See also the *OED* entry for 'bankrupt'—the Italian *banca rotta* is literally 'bank broken' or 'bench broken'.

7. J. Heyman, *Structural Analysis: A Historical Approach* (Cambridge: Cambridge University Press, 1998).

8. If we are faced with a problem with more than three forces the loop becomes a polygon with as many sides as there are forces.

9. In fact many modern bridges are built with up to three hinges placed into the arch on purpose so that the engineers know exactly where the forces are going and can arrange to resist them with suitable material and structural reinforcement.

10. H. J. Hopkins, *A Span of Bridges* (Newton Abbot, Devon: David and Charles, 1970).

11. Peter Lalor, 'Ageless, Elegant, Adored', *The Australian*, published online 10 March 2007 [online]. Available at: http://www.theaustralian. news.com.au/story/0,20867,21335574-28737,00.html.

12. Ibid.

Chapter 3: Bending It

1. 'Anping Bridge, Fujian Province', *HaloChina* [online], available at: http://holachina.com/index.php?l=en&section=routes&id=8

2. Ships—types of galley.

3. Herodotus, *The Histories* (London: Penguin, 2003).

4. You could check this out by getting a length of rubber block, or something similar, say about 50 mm by 50 mm in cross section and about a metre long. In the middle of the length of the block scribe two lines across the width of the block around the outside, say 10 mm apart. Set up the block as a simply supported beam and press down on it in the middle. You will see that the lines move apart on the bottom of the beam and move together at the top. The bottom is stretching and the top is compressing.

5. This is true for beams that have a cross section that is narrow and deep. The bending stresses along the top and bottom of very wide and shallow beams will vary. This effect is called shear lag.

6. For more detail see P. G. Sibley and A. C. Walker, 'Structural Accidents and Their Causes', *Proceedings Institution of Civil Engineers*, Part 1, 62 (May 1977). Commenting on the lack of straightness accepted by Hodgkinson they wrote: 'what in Hodgkinson's short beams was a second order effect became in the Dee Bridge of primary importance, simply because of unthinking increases in structural scale.'

7. *Report of the Commissioners appointed to Inquire into the Application of Iron to Railway Structures* (London, 1849).

8. P. R. Lewis and C. Gagg, in 'Aesthetics versus Function: The Fall of the Dee Bridge 1847', *Interdisciplinary Science Reviews* 29/2 (2004), explain the failure as metal fatigue—a phenomenon only just beginning to be identified then. The first theory about lateral torsional buckling was not published until 1899.

9. N. Rosenberg and W. G. Vicenti, *The Britannia Bridge* (Cambridge, MA: MIT Press, 1978).

10. H. J. Hopkins, *A Span of Bridges* (Newton Abbot, Devon: David and Charles, 1970).

11. Report of the Royal Commission into the Failure of the Westgate Bridge, Victoria, Australia (1971).

12. Oleg Kerensky, 'Presidential Address to the Institution of Structural Engineers, 10 October 1970, An Engineer's Ethics', *The Structural Engineer* 48/12 (Dec. 1970).

13. M. R. Horne, 'Oleg Kerensky—A Personal Tribute, 1st Oleg Kerensky Memorial Conference June 1988', *The Structural Engineer* 67/12 (June 1988).

Chapter 4: All Trussed Up

1. F. W. Robins, *The Story of the Bridge* (Birmingham: Cornish Bros, 1948).

2. I am using the word faith here to mean belief that cannot be proved to the satisfaction of everyone else. It refers therefore not just to religious belief but to the ideas and values that every individual has to hold in order to live.

3. Structurally deficient means that there are elements of the bridge that need to be monitored and/or repaired. The fact that a bridge is 'deficient' does not imply that it is likely to collapse or that it is unsafe. It means they must be monitored, inspected, and maintained.

4. Pythagoras's theorem says that the square of the hypotenuse is equal to the sum of the squares of the other two sides—so $5^2 = 3^2 + 4^2 = 9 + 16 = 25$.

5. Rather the strain is linear at first but then gradually increases until rupture. Scientists therefore define an equivalent to the elastic limit, which they call a proof stress. It is usually defined as a stress that produces a specified amount of strain—so a 1% proof stress is the stress at which the strain is 1%.

6. Ed Barna, *Covered Bridges of Vermont* (Woodstock, VT: Countryman Press, 1996).

7. An adze is a very sharp axe-like tool used to square off the tree trunks.

8. 'Smolen-Gulf Bridge', *Wikipedia*, available at: http://en.wikipedia. org/wiki/Smolen-Gulf_Bridge

9. Wilhelm Westhofen, *The Forth Bridge, Engineering*, 28 Feb. 1890.

10. This statement assumes in theory that the joints between the members are pinned, for if they were not, then the relative rotation between joined members would be restrained, thus setting up some internal bending moments. Manifestly the joints are not pinned as they have very large steel plates and rivets. The bending moments set up by these joints are unknown; however, the induced internal stresses are likely to be relatively small compared to those from the direct forces.

11. Westhofen, *Forth Bridge*.

12. Castigliano found that the displacement of any given load on a bridge, along its line of action, is equal to the rate of change of the strain energy when only that load changes and all of the other forces are held constant. He also found that the rate of change of the strain energy, when only the displacement changes, is equal to the force applied at the point of the displacement and is in the same direction. These two theorems allowed bridge builders to do a relatively simple calculation to find the reactions of a two-span beam—something they couldn't easily do before.

13. The total potential is the sum of the total strain energy and the potential of the applied loads. The theorem says that the rate of change of the total potential is zero for every possible virtual displacement, i.e. an elastic body is in equilibrium when the total potential is a maximum or a minimum. When it is maximal the equilibrium is unstable but when it is minimal then the equilibrium is stable. The total

complementary potential is the sum of the complementary energy and the potential of the reaction forces. Of all possible stress states the actual one is when the total complementary potential is a minimum. Again when it is a minimum we have stable equilibrium and maximal is unstable equilibrium.

14. The principle of virtual displacements is that a system of mass points is in equilibrium if the work done by all of the forces acting on the mass points is zero for every virtual displacement.

15. To a bridge builder detailing refers to working on all of the small items in a project which may seem trivial in overall terms but which are crucial to the success of a project and which do require constant attention to get them right.

Chapter 5: Let It All Hang Down

1. H. J. Hopkins, *A Span of Bridges* (Newton Abbot, Devon: David and Charles, 1970).

2. There is some doubt about the exact date the bridge was built. Hopkins writes (in *A Span of Bridges*) that in 1868 Charles Bender credited Finley with a bridge in 1796. Finley himself describes a bridge in 1801. The five-year discrepancy is curious. It is unlikely that Finley did not record his own work accurately. Hopkins speculates that Finley's 1796 bridge failed, and quotes as evidence that the 1801 bridge incorporated a number of details that could only have been developed from experience.

3. Theories often assume the cables are inextensible; i.e. they change in shape but don't lengthen or extend.

4. Telford proof-tested each bar to twice the estimated working stress in the bridge. During each test the bar was struck by 'some smart blows on the side with a hammer' whilst under tension and examined to see whether there were any signs of fracture. Telford's assistant, William Alexander Provis, wrote in 1828, 'It is true that their ordinates [the chains] may have been determined by calculation but with a practical man an experiment is always more simple and satisfactory than theoretical deductions' (*Menai Bridge*, London, 1828).

ENDNOTES

5. H. Seely, O. H. Ammann, N. Gray, and H. E. Wessman, 'Technical Survey—Brooklyn Bridge after Sixty Years—a Symposium', *Proceedings ASCE* 72 (Jan. 1946), 5–68.

6. Ibid.

7. Ibid.

8. Ibid.

9. Sir Isaac Newton, in England, and Gottfried Wilhelm Leibniz, in Germany, independently developed the basis of modern differential calculus, which is an indispensable tool for all modern science and technology. It is a mathematics that makes it possible to work with rates of change. Newton's Laws state that a mass will continue to move with a velocity until acted on by a force that changes the velocity, i.e. acceleration (or deceleration). The force, said Newton, is equal to the mass times the rate of change of the velocity, i.e. acceleration. In Chapter 2 we considered velocity as a vector, having size and direction. In differential calculus we think about small amounts of change of a vector quantity, such as a distance traveled, s. We express that small amount by putting a 'd' in front of s to make ds. Thus, ds signifies a small change in s. A small amount of time is dt. Velocity is the rate of change of distance, or distance divided by time. Velocity is expressed as ds/dt. Likewise acceleration is the change in velocity over a very small amount of time, i.e. dv/dt. We have said that v is ds/dt so we can substitute this into the expression for acceleration. This gives d(ds/dt)/dt, which we write as d^2s/dt^2. So if force is F and mass is m and Newton's law says that force is mass times acceleration then we can write $F = m \times d^2s/dt^2$. Now in bridge building we can express the internal forces using Newton's laws. However, instead of distance travelled by a car, we use the movement in a degree of freedom. This might be a deflection of an element of a beam or its rotation. The bending of a beam varies with its curvature— the more the curvature the more the bending. When we express the curvature using differential calculus it turns out that the bending moment in a beam is equal to the bending stiffness of the beam times the rate of change of the change of deflection—it's an acceleration, not with respect to time, but along the length of the beam. This is a differential equation that mathematicians can solve. We can produce a differential equation for a vibrating beam. The force driving the

vibration of an element of the bridge turns out to be equal to the sum of three things. The first is the mass of that element times its acceleration in a degree of freedom. The second is a damping coefficient times its velocity. The third is the stiffness of the element times the movement in a degree of freedom (such as a deflection or rotation).

10. In 1939 Richard Southwell was developing a numerical method using hand calculations, called the relaxation method. He converted the differential equations of *infinitesimal* differential calculus into a *finite* differential calculus. He concentrated on the movements in the degrees of freedom (such as deflection) at various points in a structure. In effect he created a net or mesh of points over the structure and focused on the movements at the nodes of the net. He expressed local approximations for the rates of change of the movements between those various points in terms of the movements of the surrounding points. In this way he created many linear equations which he had to solve. He did this by his relaxation method. In effect he assumed a set of movements in each degree of freedom at the points on the net. This creates a set of internal forces. However, because the assumed movements were somewhat arbitrary, the internal forces would not necessarily be in equilibrium. There is a set of residual internal forces of imbalance. If he could reduce those imbalances to zero he had solved the problem. He knew, if he changed a force, what impact that has on the surrounding movements. The techniques involved changing or relaxing the movements to bring the residual forces to zero.

11. Jacques Heyman, *The Science of Structural Engineering* (London: Imperial College Press, 1999).

12. I have referred only to the fundamental natural frequency. Both guitar strings and bridges have patterns of vibrations with higher natural frequencies that provide higher harmonics. These higher natural frequencies of a guitar string create timbre—the characteristic quality of the sound of an instrument. They are integer (i.e. whole number) multiples of the fundamental frequency. The fundamental frequency is therefore the first harmonic. If we ignore the effects of sag, a suspension cable is rather like a guitar string. However, bridges and other structures, as a whole, are much more complicated than a guitar string and the higher natural frequencies are not simple multiples of the fundamental frequency.

13. The ratios I have used are simple but very crude. Torsional stiffness is also very dependent on the form of the bridge deck (box girder/truss/plate girder, etc.) and to some extent on the cable arrangement. Long cable-stayed bridges have A-shaped pylons so the cables are in inclined planes to add to torsional stiffness. The deck girder depth could be governed by the spacing of hangers. The ratio of span to depth for bending stiffness is also very crude because although bending stiffness is proportional to width it is proportional to the depth cubed.

14. Angel of the North. http://www.gateshead.gov.uk/Leisure%20and %20Culture/attractions/Angel/Home.aspx

15. A folding structure is one that can open out or unfold from a small to a large size just as an umbrella.

Chapter 6: How Safe Is Safe Enough?

1. We can calculate the resisting moment of a beam as follows. Looking at Figure 44a we can see that the stresses change from s at the bottom to zero at the middle. The stresses are, however, the same across the width of the cross section. The force on an element is the stress times the area. So adding them all up is the same as finding the area of the triangle of stresses down the depth of the beam times the width b across the beam.

You may recall that the area of any triangle is half of its base length times its height. The base of our triangle of stresses is one half of the depth of the beam which is $(d/2)$. The height of the triangle is the stress s. So the area is one half of $(d/2)$ times s, which is $(d/4) \times s$. But the stresses also extend across the width of the beam b, so the total area is $(d/4) \times s \times b$.

We can rearrange the symbols to write

$$T = s \times b \times d/4.$$

The area of the total cross section of the beam is $b \times d$, so we'll call that A. If we can put that in our formula for T we get $T = s \times A/4$.

So what does this formula tell us? It tells us that the internal force T is one-quarter of the area of the whole beam times the stress s.

The compressive force C must be equal to T for the forces at the cut to balance.

The turning effect is the force T (or we could use C as they are the same) times the shortest distance between T and C.

Now we need to recall another fact about triangles. It is that the centre of gravity of a triangle acts a third of the way down from the top of the beam to the middle.

So the distance that C acts from the top is $(1/3) \times (d/2)$, which is $d/6$, as shown in Figure 44c. Likewise the distance that T acts from the bottom is $d/6$.

Therefore the distance between the forces T and C is $(2/3) \times d$ as shown in Figure 44c.

So the resisting moment RM is $T \times (2/3) \times d$.

Earlier we found $T = s \times b \times d/4$.

So using this we know that

$$RM = (s \times b \times d/4) \times (2/3) \times d.$$

If we rearrange the symbols we get

$$RM = s \times b \times d^2/6.$$

2. Trucks are banned and the flow of cars is controlled by a toll system.

3. There are three ways of describing an average: mean, mode, and median. For a bell-shaped curve, the so-called normal or Gaussian (after Carl Friedrich Gauss) distribution, they are numerically the same value. The mean is the sum of all values divided by the number of them. The mode is the middle value when they are all ordered in size and the median is the most common value.

4. Vitruvius, *The Ten Books on Architecture*, trans. M. H. Morgan (New York: Dover, 1960).

5. *Engineer and Contractor's Pocket Book for 1859* (London: John Weale, 1859).

6. Classical probability requires that if your belief in something is 0.3 then your belief in not that something is necessarily 0.7. There are modern open world theories of evidence that allow for you to just not know or for the evidence to be in conflict.

7. Darrell Huff, *How to Lie with Statistics* (London: Penguin, 1991).

ENDNOTES

Chapter 7: Bridges Built by People for People

1. Denis Noble, *The Music of Life* (Oxford: Oxford University Press, 2006).

2. Clearly the book metaphor breaks down when there are more than the six levels of letters, words, sentences, paragraphs, sections, and chapters. Looking inwards and downwards from letters all things are made from molecules, atoms, and subatomic particles, which according to modern physics are events and processes rather than objects. Looking outwards and upwards from whole organisms then we have teams, societies, communities, regions, nations, the whole natural environment until we get to the entire universe or even God if that is what you believe—the whole which contains all that there is.

3. This is not to be confused with the American philosophy of pragmatism which is concerned with meaning defined in terms of the practical consequences that result if something is true. I am using pragmatism as a term to describe the means by which we provide quality as fitness for purpose.

4. It is impossible to be precise since there is no official database of bridges. The estimate of 160,000 comes from the bridge owners forum at: http://www.bridgeforum.org

5. For joined up government policy see: http://www.nationalschool.gov.uk/policyhub/better_policy_making/joined_up.asp

6. Simon Caulkin, 'Why Things Fell Apart for Joined-up Thinking', *The Observer* (26 February 2006). Available at: http://www.guardian.co.uk/society/2006/feb/26/publicservices.politics

7. Ibid.

8. Michael E. McIntyre, 'Audit, Education, and Goodhart's Law; or, Taking Rigidity Seriously' (published online 2000; last updated 14 May 2002), available at: http://www.atm.damtp.cam.ac.uk/people/mem/papers/LHCE/dilnot-analysis.html

9. See the *Constructing Excellence* webpage (http://www.constructingexcellence.org.uk). This initiative is related because governments all around the world do commission and build bridges. Indeed governments are major clients of bridge builders.

ENDNOTES

10. D. I. Blockley and P. G. Godfrey, *Doing It Differently* (London: Thomas Telford, 2000).

11. Peter Senge, *The Fifth Discipline* (London: Century Business, 1990).

12. For more recent reports in the press of similar incidents, see the Victoria Climbié Foundation UK website: http://www.victoria-climbie.org.uk

13. *Ralph B Peck—Engineer, Educator, a Man of Judgement*, NGI Publication 207 (Oslo: Norwegian Geotechnical Institute, 2000).

14. Ibid.

15. Ibid.

16. John D. Barrow, *Theories of Everything: The Quest for Ultimate Explanation* (Oxford: Oxford University Press, 1990).

17. *Constructing Excellence* webpage, and Blockley and Godfrey, *Doing It Differently*.

18. See Intelligence and Security Committee, *Report into the London Terorist Attacks on 7 July 2005*: http://www.cabinetoffice.gov.uk/media/cabinetoffice/corp/assets/publications/reports/intelligence/isc_7july_report.pdf

19. The Intergovernmental Panel on Climate Change: http://www.ipcc.ch/

Glossary[*]

Bending moment   The internal force in a beam generated when two ends of an element of material are rotated in opposite directions.

Caisson   A large water-tight case or chamber normally built to enable dry working on a foundation. It may be built partly or wholly above ground and sunk below ground usually by digging out the soil inside.

Camber   The slight upward curvature of a bridge deck.

Cantilever   A beam that overhangs—fixed at one end and free at the other—like a diving board.

Catenary   The curve formed by a flexible cable as it hangs under its own weight from two points.

Centring   A temporary framework to support an arch during construction.

Corbel wall   An opening in a wall built by progressively projecting the bricks or stones out a little more at each layer.

Creep   A gradual increase of strain and deformation of a material when under a constant load or stress.

Damping   The process by which a bridge loses energy due to friction and a general 'looseness' between the parts of the bridge.

Degrees of freedom   The number of independent coordinates needed to describe the configuration of a bridge.

Elastic   The ability of a material to regain its original shape after being changed—for example, by being pulled or compressed.

Elastic modulus   The ratio of stress to strain for an elastic material.

---

[*] With acknowledgement to D. I. Blockley, *The New Dictionary of Civil Engineering* (London: Penguin, 2005).

GLOSSARY

Elastomeric laminate Layers of steel plate encased in an elastomer polymer, rubber, or synthetic rubber material—used in a bridge bearing.

Encastré support A support that constrains all movement—also known as a fixed support.

Energy A capacity for doing work. Technically, the main forms of energy are kinetic (due to the motion) and potential (due to position relative to a datum). Strain energy is a form of potential energy.

Finite element analysis A computer-based method of analysing the behaviour of structures. It relies on splitting the structure up into a large number of discrete 'bits' or elements. Equations relating the internal and external forces and displacements for all the elements are set up and solved in a way that would be difficult to do manually.

Fixed support A support that constrains all movement—also known as an encastré support.

Flange The wide parts, top and bottom, of an I-beam.

Flutter A form of wind-excited oscillation.

Force That which is required to accelerate a mass—measured in newtons.

Hogging Bending upwards to form a shape that is concave below—the opposite of sagging.

Kentledge Any heavy material, e.g. large blocks—used to give weight and stability.

Metal fatigue The deterioration of the strength and other mechanical properties of a material when subjected to repeated applied stress in the elastic range. A normally ductile material may become brittle and crack.

Natural frequency The frequency of free vibration of a system.

Neutral axis The axis of a cross section of a beam where the stress due to bending is zero.

Newton The force which accelerates a mass of 1 kg by 1 m/sec/sec.

Pile A column-like structural member of concrete, steel, or timber driven, jacked, or cast in the ground to support the structure above ground.

GLOSSARY

Pinned support   A support where the structural members being joined can rotate freely with respect to each other.

Plastic   Where a material does not regain its original shape after being pulled, compressed, or sheared. Plastic deformation is the change not recovered when the load has been removed.

Point of contraflexure   Where the bending of a beam changes direction, e.g. from hogging to sagging. At this point the bending moment is zero.

Prestressed concrete   Concrete in which stress is induced during the manufacturing process to increase eventual strength.

Reinforced concrete   Concrete with steel reinforcement needed because concrete is weak in tension.

Resonance   Where a bridge has very large vibrations—occurs when the frequency of an exciting force coincides with the natural frequency of the system.

Rivet   A metal rod used to join two pieces of metal together. The rivet, cold or hot, is inserted through aligned holes, and the head at one end is held tightly while the projecting end is hammered to form another head, thus gripping the material.

Sagging   Bending downward to form a shape that is concave above—the opposite of hogging.

Scalar   A quantity with magnitude but not direction—cf. vector.

Shear force   An internal force generated when layers of material are prevented from sliding over each other.

Simple support   A support that provides only a vertical reaction.

Spandrel   The roughly triangular area between the outer curve of an arch and the bridge deck.

Statically determinate   Where the internal forces in a bridge can be found by balancing the forces (simple statics) alone.

Stiffness   The relationship between load and deformation of a structural member.

Strain   The deformation produced by a load.

Strain energy The potential energy stored in a structural member when it is elastically strained.

Stress The force on a defined area of material.

Strut A structural member in compression.

Tie A structural member in tension.

Vector A quantity with direction as well as magnitude.

Voussoir A wedge-shaped block in a masonry arch.

Web The vertical part of an I-beam that joins the flanges.

Weld The process or product of joining materials by heating the surfaces of the pieces to be joined to fuse or coalesce them using pressure or a molten filler to make the joint.

Work Just as human work is a physical or mental effort so the technical definition of work is the effort of a force to move a distance—expressed as the force multiplied by the distance.

Bibliography

Angier, Natalie, *The Canon: The Beautiful Basics of Science* (London: Faber and Faber, 2008).

Aparicio, A. C., and J. R. Casas, 'The Alamillo Cable Stayed Bridge: Special Issues Faced in the Analysis and Construction', *Proceedings of the ICE—Structures and Buildings* 122 (Nov 1997), 432–50.

Barna, E., *Covered Bridges of Vermont* (Woodstock, VT: Countryman Press, 1996).

Barrow, J. D., *Theories of Everything: The Quest for Ultimate Explanation* (Oxford: Oxford University Press, 1990).

Beldon Scott, J., 'The Art of the Painter's Scaffold, Pietro da Cortona in the Barberini Salone', *Burlington Magazine* 135/1082 (May 1993), 327–37.

Blockley, D. I., *The Nature of Structural Design and Safety* (Chichester: Ellis Horwood, 1980).

—— and P. G. Godfrey, *Doing it Differently* (London: Thomas Telford, 2000).

Boer, Reint de, *The Engineer and the Scandal* (Springer, 2005).

Casas, J. R., 'A Combined Method for Measuring Cable Forces: The Cable Stayed Alamillo Bridge, Spain', *Structural Engineering International* 4 (1994), 235–40.

Chastel, A., et al., *The Sistine Chapel* (London: Muller, Blond and White, 1986).

Cossons, Neil, and Barrie Trinder, *The Iron Bridge* (Moonraker Press, 1979).

Dallard, P., T. Fitzpatrick, A. Flint, S. Le Bourva, A. Low, R. Ridsdill Smith, and M. Willford, 'The London Millennium Footbridge', *The Structural Engineer* 79/22 (20 Nov 2001).

——, ——, ——, A. Low, R. Ridsdill Smith, M. Willford, and M. Roche, 'London Millennium Bridge: Pedestrian-Induced Vibration', *ASCE Journal of Bridge Engineering* (Nov/Dec 2001).

BIBLIOGRAPHY

De Bono, E., *Practical Thinking* (London: Penguin Books, 1971).

Ekeberg, P. K., and K. Søyland, 'Flisa Bridge, Norway—A Record-Breaking Timber Bridge', *Proceedings of ICE—Bridge Engineering* 158 (2005), March Issue BE1, 1–7.

'Santiago Calatrava', *El Croquis Editorial* (Madrid: El Croquis, 1994).

Fitzpatrick, T., P. Dallard, S. Le Bourva, A. Low, R. Ridsdill Smith, and M. J. Willford, *Linking London: The Millennium Bridge* (London: Royal Academy of Engineering, 2001).

Gigerenza, G., *Reckoning with Risk* (London: Penguin Books, 2002).

Gimsing, N. J., Cable Supported Bridges (Chichester: Wiley, 1997).

Goodman, R. E., *Karl Terzaghi* (Reston, VA: American Society of Civil Engineers, 1999).

Herodotus, *The Histories* (London: Penguin, 2003).

Heyman, J., *The Science of Structural Engineering* (London: Imperial College Press, 1999).

Hopkins, H. J., *A Span of Bridges* (Newton Abbot, Devon: David and Charles, 1970).

Huff, D., *How to Lie with Statistics* (London: Penguin Books, 1991).

Jodidio, P., *Santiago Calatrava* (Taschen, 1998).

King, Ross, *Michelangelo and the Pope's Ceiling* (London: Pimlico, 2006).

Koestler, A., *The Ghost in the Machine* (London: Picador, 1967).

Lewis, P. R., and C. Gagg, 'Aesthetics versus Function: The Fall of the Dee Bridge 1847', *Interdisciplinary Science Reviews* 29/2 (2004).

Macdonald, J. H. G., 2008, 'Pedestrian-Induced Vibrations of the Clifton Suspension Bridge, UK', *Proceedings ICE—Bridge Engineering* 161 (June 2008) Issue BE2, 69–77.

McIntyre, Michael E., 'Audit, Education, and Goodhart's Law, Taking Rigidity Seriously', 2000 [online]. Available at: http://www.atm. damtp.cam.ac.uk/people/mem/papers/LHCE/dilnot-analysis.html

Melbourne, C. (ed.), *Arch Bridges* (London: Thomas Telford, 1995).

Noble, Denis, *The Music of Life* (Oxford: Oxford University Press, 2006).

Paxton, R., *100 Years of the Forth Bridge* (London: Thomas Telford, 1990).

Popper, K. R., *The Logic of Scientific Discovery* (London: Hutchinson, 1959).

Pugsley, A., *The Theory of Suspension Bridges* (London: E. Arnold, 1957).

——, *The Safety of Structures* (London: E. Arnold, 1966).

'Ralph B Peck, Engineer, Educator, A Man of Judgement', Publication 207 (Oslo: Norwegian Geotechnical Institute, 2000).

BIBLIOGRAPHY

Robins, F. W., *The Story of the Bridge* (Birmingham: Cornish Bros, 1948).

Rosenberg, N., and W. G. Vicenti, *The Britannia Bridge* (Cambridge, MA: MIT Press, 1978).

Royal Commission, Report of the Royal Commission into the Failure of the Westgate Bridge, Victoria, Australia, 1971.

Ruddock, Ted, *Arch Bridges and Their Builders 1735–1835* (Cambridge: Cambridge University Press, 1979).

Saint, A., *Architect and Engineer* (New Haven, CT: Yale University Press, 2007).

Seely, H., O. H. Ammann, N. Gray, and H. E. Wessman, 'Technical Survey—Brooklyn Bridge after Sixty Years—a Symposium', *Proceedings ASCE* 72 (Jan. 1946), 5–68.

Senge, Peter, *The Fifth Discipline* (London: Century Business, 1990).

Shiner, L. E., *The Invention of Art* (Chicago: University of Chicago Press, 2001).

Sibley, P. G., and A. C. Walker, 'Structural Accidents and Their Causes', *Proceedings ICE* Part 1 (1977), 62.

Sleeper, R. W., *The Necessity of Pragmatism* (Champaign: University of Illinois Press, 1986).

Troyano, L. F., *Bridge Engineering* (London: Thomas Telford, 2003).

Virlogeux, M., 'Bridges with Multiple Cable-Stayed Spans', *Structural Engineering International* 11/1 (2001), 61–82.

——, C. Servant, J. M. Cremer, J. P. Martin, and M. Buonomo, 'Millau Viaduct, France', *Structural Engineering International* 15/1 (2005), 4–7.

Watanabe, Y., 'Miho Museum Bridge, Shigaraki, Japan', *Structural Engineering International* 12/4 (2002), 245–7.

Westhoven, W., *The Forth Bridge* (London: Engineering, 1890).

Index

INDEX

INDEX

INDEX

INDEX

INDEX

Lightning Source UK Ltd.
Milton Keynes UK
UKOW06f1137050616

275589UK00003B/6/P